DAMS
Incidents and Accidents

DAMS

Incidents and Accidents

K.R. Saxena
V.M. Sharma

CRC Press
Taylor & Francis Group
Boca Raton London New York

CRC Press is an imprint of the
Taylor & Francis Group, an **informa** business
A TAYLOR & FRANCIS BOOK

First published 2005 by Taylor & Francis

Published 2022 by CRC Press
Taylor & Francis Group
6000 Broken Sound Parkway NW, Suite 300
Boca Raton, FL 33487-2742

Visit the Taylor & Francis Web site at
http://www.taylorandfrancis.com

and the CRC Press Web site at
http://www.crcpress.com

British Library Cataloguing in Publication Data
A catalogue record for this book is available from the British Library

Library of Congress Cataloging in Publication Data
A catalog record for this book has been requested

Publisher's Note
The publisher has gone to great lengths to ensure the quality of this reprint
but points out that some imperfections in the original may be apparent

Foreword

Designed and constructed man-made structures require maintenance and repair or rehabilitation to prevent accidents. Nature is more powerful than human knowledge and unexpected or unknown factors, particularly environmental, cause catastrophic incidents. Of all structures, dams holding water reservoirs are most prone to such dangers. With increasing heights of dams and volume of reservoirs, particularly at less favourable sites, problems have been further compounded. Additionally, the durability of materials used in construction of dams has its own limitations. Chemical reactions among their constituents may take place subsequent to construction. No factor of safety of any order can compensate these situations.

Saxena and Sharma, experienced engineers of repute, whom I happen to know personally, have attempted this difficult but useful topic for their book. The limitations or misconceptions of design have been presented, suggesting methods of investigating dam incidents and accidents and pleading an unbiased approach to take the knowledge forward. It is very true, as the authors have illustrated, that it is difficult to reproduce the conditions prevailing at the time of incidents or prior to failure of a dam. This creates controversies about the causes of dam incidents and accidents, as a number of agencies, owners, investigators, designers, and construction engineers, have to safeguard their own interests. Human knowledge is dynamic and dam incidents need investigation in the background of the existing status of knowledge, to find out the real cause of incidents. To learn from the dam failures of the past, it is worthwhile that select dam failures are re-examined in the light of current knowledge.

The book is written in lucid language and describes the events of dam incidents and accidents. It is a useful addition to the literature on dam engineering. I hope practicing engineers and those concerned with dam construction and associated post-construction problems, will find it a useful reference book.

May 16, 2004
New Delhi

C. V. J. Varma
Honorary President
International Commission on Large Dams

Preface

One of the most outstanding and prestigious activities of civil engineering profession has been construction of dams, which symbolize the vast development made by humankind, its progress and prosperity. The increase in the number of dams, their heights and lengths, over the last century, represent the race between increase in population and demands of food, power and industry. The reservoirs store rainwater and absorb floods, thus saving us from unabated misery, loss of human life and property, a situation man has been facing for a long time. Farmers use water, where and when needed; people tap power from grids running across countries; factories produce goods in quantity and quality, required for needs and comforts of a large population; all of these have contributed to the revolution in human civilization.

Dam construction, normally across a river, is based on its conceptual behavior, generally covered by the term 'design'. Design is based on the hydraulics of water and other basic elements responsible for dam performance—the durability and engineering properties of materials used in the structure and their interaction with other materials, including water; the site geology as understood or gathered by an engineer and/or engineering geologist. Besides these issues, the technique and skill put in the construction of a dam and how best it is supervised at the first filling stage and thereafter maintained during its service life, all contribute to its health.

Engineering theories, when practiced on the ground, enrich one's experience in the standard procedures that have been followed under particular natural conditions. An experienced design engineer is aware of the fact that every theory has certain assumptions, with unknown or calculated risks, and is applicable only under certain conditions. Efforts are made to minimize these risks, maintaining a possible balance between assumptions and economy, as the vagaries of nature, which normally cause havoc, are highly unpredictable. Applications of the same theory in an extended form is the engineer's professional ingenuity and depends on how best the basic assumptions have been rationalized, without infringing on the limitations. Every site has its own problems for a dam designer and type designs or standard procedures do require modifications accordingly. Missing or underestimating any existing pronounced site conditions, geological or geomorphological and rainfall or runoff data usually cause disaster at any stage. It is very often mentioned that the 'site was good for a reservoir but not for a dam', but the truth rarely comes out. The case histories included in this book the highlight these underlying truths, which is primary requirement of an investigation of any incident. Any repair or rehabilitation work—its necessity and economy, not discussed in the book—is secondary.

The review presented in the book shows that the dams constructed in earlier periods in different countries totally depended upon experiences and unexpected/unpredictable floods, which had been the main cause of their deterioration and failure. The understanding of the engineering properties of materials and development of methods for their determination made a great impact on their use and placement in dams. With the advent of rock mechanics and engineering geology, the foundations of dams became stronger. Seepage control and improved drainage measures made the foundations an integral part of the dam.

The most important addition to the observational methods has been continuous monitoring and surveillance of dam with the use of mechanical, electrical or electronic instruments, judiciously installed in the body of the dam. But the success of these methods depended upon their sensitiveness, quick response and immediate analyses of data. Unfortunately, all these methods—as is experienced—may not rule out disaster or failure of the dam. The warning recorded by the first sign of a dam incident may not give enough time for improvement work as some other system, like lifting of gates over the spillway to lower the water level in a reservoir in a controlled manner, may not work. Contrived and controlled failures can teach lessons, but this is rarely possible, except that a good number of lives and movable properties are saved.

The cases of failures of dams prove that there have always been controversies in searching for the reasons of a failure. One of the reasons for this is that there is not a single reason of failure and the signatures left behind after a disaster are too many and too complex. Lessons are learnt, but removal of causes of failure, which become an inherent part of a dam or reservoir, remain a remote possibility. All rehabilitation measures aim at acknowledging the weaknesses exposed and designing dams accordingly, leaving the compatibility of old and new cross-sections to the construction engineers.

Besides the case histories of incidents and disasters of dams, the other chapters in the book deal with the misconception in design, methods of investigations of dam incidents. The aims and objectives behind the study of failures, are complementary and do not advocate anything specific. As has been said, 'little is learnt from an unfailed dam but failures cannot be afforded'. Dam incidents and accidents, generally known as deterioration of dams, need greater attention. Once the potential risk is eminent, efforts are needed to make failures as 'contrived and controlled' as possible, with the aim to learn more than what is seen at the end.

The authors gratefully acknowledge the technical material made available by Dr A.D.M. Penman, Geotechnical Engineering Consultant, UK, and the Central Water and Power Research Station, Pune, India, for providing reference facilities of the library, which helped to write this book. The assistance of Akshay Sahai and Shrey Vardhan in computerizing the data and proofreading is also appreciated.

July 2004

K.R. Saxena
V.M. Sharma

Contents

Chapter 1

Introduction

Life, as we know it on our planet, began in water, evolving slowly and spreading out to inhabit most of the world's surface. All living things contain and require water for their continuing existence. Seeds can exist for several years in an arid desert, yet burst into life with the first shower of rain and carpet the desert with beautifully coloured blooms. Each bloom then does its utmost to ensure that its pollen will reach another flower so that there will be more seeds to wait in the hot sand until they, in turn, can generate life with the first drop of water.

The exact number of species that inhabit our earth is not known correctly, but has been estimated by Morris (1998) at twenty million, with even more that once having existed and now become extinct. Each species is doing its best to reproduce so that it may dominate the world, but control comes through the species preying on each other. After trillions of years of evolution, man has evolved into the dominant species and is busy reproducing at an alarming rate so as to fill the whole world and control all other species. Now, with our numbers exceeding six billion, our need for fresh water for ourselves, the animals we rear for food and the crops we grow, is testing the limits of all available supplies. Many rivers pour their fresh water into the saline ocean body and to survive, we have to arrange for as much storage of this precious water as possible.

WATER POWER

A person's attitude towards water is likely to depend on individual experience. If you have been swept off your feet by flood and carried out into a fast-flowing river, clinging on to floating debris and hoping to be rescued before you were dashed against rocks or sucked into a rapid, your view of water power is likely to be very different from someone who has played as a child in a babbling brook, perhaps learning to swim in a still pool, later lying in the lush meadow grass beside the same brook while his or her own children play in the same place. If you had to carry water from a well or stream to your home, you will have a very different assessment of the value of water as compared with a person who has always had water piped to their house and expects cool, clean water to flow from a tap every time it is turned on. People living on a small holding on fertile ground near a river may not see the need nor advantage in storing water if their land is to become the bottom of a reservoir. Bureaucratic demands may require them to move to a much less attractive site, but they know that it may be many years before their land goes under water, and they may be able to get several crops gathered in before their house is submerged. During that interval, there is time for organized protests to develop against the construction of the dam, and those unable to see the overall benefit to be gained from the storage of water may put forth strong arguments on behalf of the many small holders in the valley, stating that they should be left to continue with their lifestyle that may be many generations old. One argument is that while the proposed storage will be a huge benefit to the population as a whole, the displaced small holders may experience little or no benefit at all. There is no doubt that storage of potable water is a vital necessity for the improvement of our civilization, but even with completed schemes, there is the risk for those living downstream of a dam, that its failure could lead to dangerous flooding, and it is this aspect—dam failure—that is the subject of this book.

As we all know, water is a transparent liquid of low viscosity that is extremely incompressible. Water has been used as a hydraulic fluid to transmit pressure for the operation of presses, lifts and cranes. The pressure developed by a hand pump, when fed to a piston of a large diameter, can produce a force sufficient to shape metals, uproot trees, lift bridges, etc. In some cities, companies were formed that supplied water under pressure that was distributed through pipes to buildings for the operation of their lifts. Such a system was used in London docks for the operation of their cranes, loading and unloading ships. Hydraulic pressure caused by the depth of water in reservoirs acts within and under dams, and can be the cause of damage and, in extreme cases, of failure.

STORAGE OF WATER

The need to store water is rooted in the beginnings of our civilization. When humankind began growing food rather than just gathering it and hunting—a change brought about by the greater number of people and a desire for a higher standard of living—it soon became necessary to store water for irrigation. Dams and their remains have been found in the recognized centres of early civilizations, in Egypt, Mesopotamia, India, South America, China, etc.

Increasing population led to a proportionate demand for more dams to supply water, not only for irrigation purposes but also for the increasing numbers and sizes of towns and also to the power mills and sawbenches. Industrialization put ever greater demands on water supply and the development of hydroelectric power in the late nineteenth century was welcomed as the solution to mankind's demand for energy, in much the same way as nuclear power was feted in the twentieth century.

Dams are an inevitable feature of civilization as rains are confined to certain period of the year, maybe for 100 hours spread across 100 days, whereas the requirement of water for various purposes is required for 365 days. The cycle of requirement not only remains constant but increases due to various developmental activities directly related with increase in population. But unfortunately, nature may not provide the same quantity of water every year, or even more to meet the extra requirements. Hence, water has to be stored in order to supplement the rainfall shortage, if any, in subsequent years. This adds to the height of dams to increase storage, as can be seen these days. Add to this, rainfall is skewed geographically and average. In the case of India, varies from 100 mm in the west to 1,170 mm in the east. For inhabitants of Cherrapunji in the east, which gets highest rainfall in the world annually, water is sold via tankers for lack of storage after the rainy season and in the west, they have to depend on water trains for their every day needs.

The storage of large volumes of water retained by dams, although does provide great benefits, it also constitute a risk, particularly to those living downstream. The aim of this book is to briefly highlight the self-evident benefits derived from the very large number of dams to be found worldwide and give examples of some of the failures that have occurred, with the resulting damage and death. With steady improvement in the art and science of dam design and construction, it has become possible to build dams ever higher, thereby storing water more efficiently and also to be much safer against the risk of failure. We may have reached the stage when dams can be built not to fail, yet some failures continue to occur, often due to preventable faults. No two dam sites are the same, so there can be no one specific standard dam. Every reservoir has to be designed as an individual and its success is dependant firstly on the skill of the designers and constructors, then subsequently on correct and timely maintenance. A dam with adequate maintenance will last forever. Many the early dams only failed after the society they served had been destroyed, dispersed or for some reason, no longer needed the dam. An attempt has been made, in the preparation of this book, to offer some explanation for the failures described, for the interest and information of the reader.

Chapter 2

DAMS—Incidents and Failures

For centuries dams have provided mankind with such essential benefits as water supply, flood control, hydropower, irrigation, recreation, etc. Dam failures are rated as one of the major 'low probability high loss' events. Many of the older dams are characterized by increased hazard potential due to the development downstream and increased risk due to structural deterioration and inadequate spillway capacity. The International Commission on Large Dams (ICOLD) have published a World Register of Dams (1998) which gives some facts regarding the number of dams in the world and their present status.

There are 25,410 dams over 15m high, of which 12,000 have been built for irrigation, 6.500 for hydropower and 5,500 for water supply, although many of them serve more than one purpose. ICOLD member countries have reported 41,413 operational dams in 1996.

The demand for water is increasing steadily with corresponding growth of human population. This demand is estimated to have risen seven times between 1900 and 1995, more than twice the population growth rate of 90 million/year.

Rainfall across all the world's land amounts to 110,000 km^3/year. Of this, 70,000 km^3 evaporates and is used by plants, 14,000 km^3 goes into the rivers, 26,000 km^3 soaks into the ground, meaning that of the total, 40,000 km^3 runs into the oceans. The aggregate storage of all dams is 6,000 km^3 and, as claimed by the Position Paper on Dams and the Environment, more dams are needed to fill the existing void.

There are 18,000 single purpose dams, comprising 48% irrigation, 20% hydroelectric power, 15% water supply, 8% flood control and 4% recreation.

Multipurpose dams number 7,400, i.e., 30% of the total. Of these, 48% serve the purpose of irrigation, 39% towards flood control, 36% hydroelectric power, 36% domestic and industrial supply, 24% recreational and 5% for fish farming and navigation. Embankment dams of earthfill predominate over the others, forming 64% of all reported dams. Those of rockfill form 8% and the world's highest dam, the 300m high Nurek Dam in Tajikistan, is an embankment type dam, although most are less than 60m high. Masonry or concrete gravity dams form 19%, arch dams 4%, buttress dams 1.4%, barrages 0.8% and multiple arch dams 0.5%.

Less than 30m height form 62% of the reported dams, while those less than 59m form 90% and those over 100m form just over 2% of the total number of dams world wide.

The total number of dams over 15m high in 1950 was 5,749, while by 1990 the figure had raised to 23,112, i.e. a fourfold increase in 40 years.

The world's highest dam would have been the 335m high Rogun dam in Tajikistan that had been designed for hydroelectric power and irrigation and begun in 1988, but the capitalistic success in smashing Russia brought work to a stop, leaving the 300m Nurek embankment dam, completed in 1980, to remain the highest Dam till date. Next in height order is the 285m high Grande Dixence Dam in Switzerland, built for hydroelectric power, in 1961. It is followed by the 272m high Inguri Dam in Georgia (Europe), built for hydropower in 1980. Next is the 262m high Vaiont Dam in Italy, also for hydropower, built in 1961.

Of all the dams built, 6,375 have been constructed in the USA, 4,010 in India, 1,855 in China, 1,187 in Spain and 1,077 in Japan.

Table 2.1 Number of dams built in the period 1900-1989

1900 – 1949	less than 1000 in every 10 years
1950 – 1959	2733
1960 – 1969	4788
1970 – 1979	5415
1980 – 1989	4427

In terms of reservoir capacity, Kariba Dam tops at 180,600 × 10^6 m^3, High Aswan holds 162,000 × 10^6 m^3 water, and the Three Gorges, when completed, will store 39,300 × 10^6 m^3.

Some of the dams were built in ancient times. The table below lists the years in which they were completed, with the birth of Christ as zero.

Table 2.2 Earliest dams constructed but still in use

– 543	Khlongkathun for irrigation and water supply in Thailand
+ 100	Cornalbo for irrigation in Spain
+ 100	Proserpina for irrigation, water supply and recreation
+ 130	El Bercial for irrigation in Spain
+ 700	Mannoike for irrigation in Japan
+ 1000	Thonnur Tank for irrigation in India
+ 1128	Daimonike for irrigation in Japan

Table 2.3 Recorded remains of dams built in the period Before Christ (BC)

– 3100	Kosheish on the Nile near Memphis: the ruins still exist.
– 2500	Atheim, a masonry dam on the Atheim river, a tributary of the Tigris.
– 800	Sennacheirib's dam of stone construction on the Tigris in North Nineveh.
– 700	Atrush, a masonry dam on the Atrush river, a tributary of the Tigris, north of Mosul.
– 500	Diyala masonry dam on the Tigris, north of Baghdad.
– 500	Abu Habea masonry dam, also in Iraq.
– 240	Gukow rock 30m high caissons in China.

In 1980, Dr Londe spoke on the lessons from Earth Dam Failures at the Asian Institute of Technology. He described the history of dam building since the dawn of civilization as a long series of failures. Man learned little from success but a lot from his mistakes. The failure of a dam was, however, the result of a complex concourse of causes and mechanisms which should be interpreted with extreme care. He gave several case histories of earth dam failures and arranged them in three groups: failure by sliding, failure by overspilling and failure due to water seepage. Each group led to specific lessons and Dr Londe pointed out how learning from our errors is vital for improving our knowledge and promoting a safer design. Designing and building a successful earth dam is even now more an art than a science. In the light of these lessons, Dr Londe put forward ten recommendations for improving the overall safety in the design, construction and operation of earth dams. These were: (1) Look out for over-consolidated clay formations, and use residual strength parameters; (2) Look out for loose saturated sand formations and study their liquefaction potential; (3) Analyse the flood in terms of probability of occurrence and corresponding probability of damage downstream; (4) Use the most recent hydrological methods; (5) Give detailed and strict operating rules for gated spillways; (6) Provide ample and well-graded filters and drains to prevent piping; (7) Test for dispersion potential of fine clayey soils; (8) Incorporate instrumentation as a vital part in the design of a dam for monitoring its safety; (9) Make use of inspection galleries for instrumentation and for quick remedial action, if required; and (10) Provide a thorough and careful surveillance of dams and appurtenant structures on a continuing basis. In the cases studied in this book, we will be able to plumb the depth indeed of these recommendations.

While earth dams are the most numerous in our world population of dams, embankment dams have also been built of rocks and there are many combinations that have been used of rocks, boulders, sands and gravels, reducing the size of particles through silt and silty clay to clay.

All of these materials can be won from the surface of the world, and it is usually best to use the materials that can be found at the site for the dam.

In the past, quarried rock has been shaped to pieces suitable for masonry construction of dams. With the development of hydraulic cement made from limestone and clay, the sand and gravel fraction of excavated earth can be used to make concrete. During the nineteenth century, concrete technology and methods of structural analysis gradually evolved. Combined with the fact that at that time, there were many sites available offering sound rock foundations at shallow depths, an increasing interest developed for the use of concrete dams.

This trend was countered during the first quarter of the twentieth century by the development of soil mechanics which gave an increasing understanding of the behaviour of embankment dams.

EARLY DAMS

Embankment dams have been built for the use of man since the fourth millennium, BC. Physical evidence of them may have been lost and our knowledge of their existence comes mainly from written records. According to Van Loon (1922), mankind's recorded history of dam construction began about 6,000 years BC in Egypt. An advanced civilization had developed which was skilled in irrigation methods. In the rest of the world, the early role of dams in providing for irrigation water formed a major contribution to the development of civilization. The oldest dam in the world, according to Kerisel (1985), dates from around 4000 BC and was built of earth with a masonry facing at Jawa in Jordan. Notable water supply and irrigation systems have existed from the fourth millenium BC. Geographical locations where major systems have been developed include Egypt, Mesopotamia, Cambodia, China, Greece, Rome, India, Mexico, Palestine, South America and Urartu. Pearson et al; (1982) have traced such schemes over the period 5000 BC to 1200 AD and state that many of the schemes persisted for thousands of years after their inception.

Egypt (5000 BC to 1800 AD)

Large dams were not needed for irrigation in the Nile valley, but many small earthen dykes and irrigation canals were built in many instances. It is in Egypt that we find the remains of one of the oldest known dams, the Sadd-el-Kafara dam, built, according to Smith (1971), during the period 2950 to 2750 BC. It was designed to retain a reservoir in the Wadi el-Garawi, near Helwan, 30km south of Cairo. As indicated by Fig. 2.1, it was 11.3m high with a crest length of 106m, and consisted of two rubble

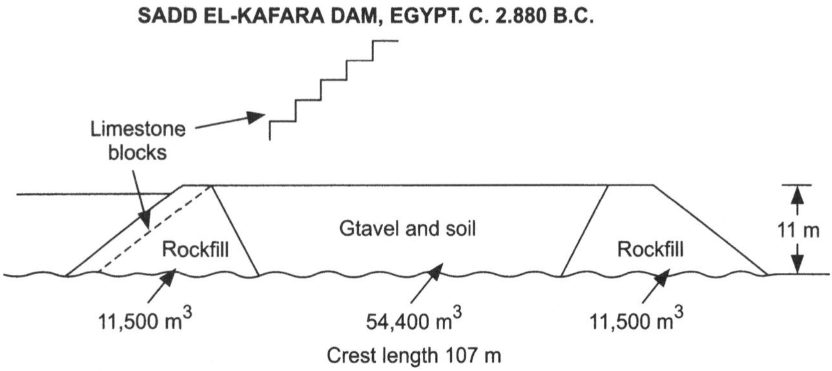

Fig. 2.1 Section of Sadd-el-Kafara dam, Egypt

masonry shoulders supporting a 36m thick central "core" of sand and gravel. The total thickness at the major section was 84m at the base, with a crest width of 61m and up and downstream slopes of 1 on 1. The upstream slope was finished in limestone blocks to form shallow steps placed without mortar, presumably as wave protection.

The reservoir area at the Sadd-el-Kafara dam contains very little silt, so it is presumed that it was not used for a very long period before failure occurred. The catchment area is more than 180 km^2, and records taken during the early twentieth century show that in 3 years out of 4, rainfall of more than 10mm may have occurred in one day, and that during the fourth year, it may have been 20mm a day. The reservoir could be filled by a rainfall of 20mm, so it is probable that it would have become full within a few years. Failure appears to have been caused by overtopping, producing a central breach 45m wide. After the collapse, there appears to have been no attempt at repair. It was the only dam in Egypt until modern times, which suggests that either its construction was regarded as an experiment not worth repeating or that it had aroused the wrath of the gods who had destroyed it and might do worse "if man continued with such ungodly ways".

Mesopotamia (4000 BC to 1200 AD)

Major irrigation works and structures that pre-date other schemes are to be found in Mesopotamia. It can be said that Mesopotamia was developed by a series of dams and canals, and quite substantial construction was required at canal headworks. Reservoirs were built to store water for release during the dry season and traces of irrigation canals dating from 5000 BC have been found at Choga Mami, a prehistoric site near Baghdad. The early kings of Sumer constructed irrigation works around 3000 BC and construction continued during the Babylonian Empire. The well-known Code of King Hammurabi (1800 BC) contains Section 53 relating to dams. The translation (Smith 1971) reads, "If anyone be too lazy to keep his dam in proper condition and does not keep it so, if then the dam breaks and all the fields are flooded, then shall he in whose dam the break occurred, be sold for money and the money shall replace the corn which he has caused to be ruined".

The Nimrod dam was built across the Tigris north of Baghdad, during the second millennium BC in order to divert the river to another course. As with all dams, some maintenance is required if they are to survive and the failure of this dam in about 1200 AD may be associated with the decline of governmental power and authority. Agriculture was suffering from salination and a general decay had set in before Hulagu Khan and his Mongol hordes swept through in 1258 AD.

China (780 BC to 1368 AD)

There are numerous historical accounts of the construction and use of irrigation systems in early China from 780 BC up to the end of the Yuan Dynasty in 1368 AD. Together with irrigation, these systems involved large-scale works for flood control and canals for transport. Little physical evidence remains of these systems, probably because of intense constructional activities and agriculture carried out in China during the intervening centuries.

Mexico (150 BC to 1500 AD)

Civilization emerged in the lowlands of the Gulf of Mexico as early as 1000 BC reaching a climax with the great cities of Teotihuacan and Tenochtitlan. The highly developed civilization of the period relied heavily on irrigation systems for agriculture. One of their bigger dams was the earth fill dam Purron, constructed in five phases during a period of 750 BC to 1500 AD. It had a maximum height of 18m, an embankment of more than 2½ millions m^3 and impounded a reservoir able to irrigate approximately 700 hectares. The transverse and longitudinal section of the dam has been given by Benson et al; (1983) from which Figure 2.2 has been prepared. No details of the spillways are known accurately and it is presumed that the dam failed from overtopping after it had been abandoned.

Urartu (850 to 600 BC)

The ancient kingdom of Urartu existed where the current borders of Turkey, Iran and the former Soviet Union meet. During the reign of 13 Urartu kings between 850 and 600 BC, the kingdom rose to a leading position. The first king was Arame, who gave his name to the Armenians,

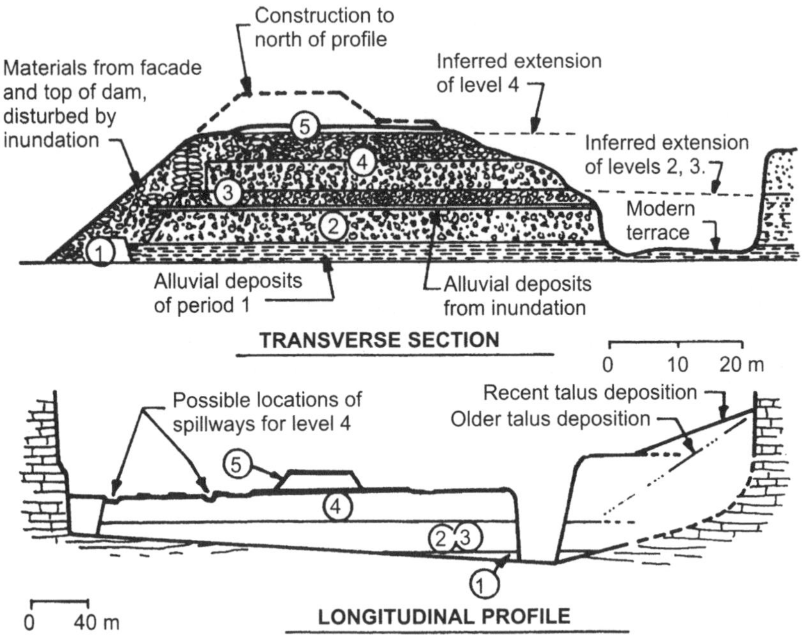

Fig. 2.2 Purron Dam in Mexico, showing the five construction phases, 750 B.C. to 1500 A.D. (after Benson et al, 1983)

present day descendants of the Urartians who were masters of hydraulic installations. Evidence of their reservoirs, dams and irrigation works can be found throughout the region. One major project, the creation of Lake Rusa during the period 730 to 645 BC required damming the natural outlets of the large Kesis Golu basin to provide the irrigation water supply for the whole region. Lake Rusa was in use for about 2500 years until the original North dam failed in 1891 AD as a result of lack of maintenance along with unusual floods. It was found that the ancient outlet through the South dam now had been allowed to become blocked, leading to overtopping of the North dam. It has now been rebuilt, putting Lake Rusa back into use.

Roman Empire (300 BC to 600 AD)

From the point of view of dam building, the Romans learnt from those they conquered. As their empire spread into the Middle East, they both inherited the existing irrigation schemes and learnt the dam designs of the Nabataean, who used twin vertical masonry walls sandwiching a core of lime rubble, supported against the pressure of the reservoir water by an earth fill shoulder on the downstream side. According to Vogel (1993), the Nabataean provided a gravel drainage between the masonry wall and the earth fill in order to reduce the pressure from the earth fill when the reservoir was drawn down; a design that they incorrectly copied later in Spain. In the near east, the new Roman garrisons required a steady water supply, and to produce a suitable reservoir, during the late second century AD, they built a 6m high masonry dam at Qasr Khubbaz, 30 miles west of Hit on the Euphrates. This reservoir has a vertical upstream side and is stepped on the downstream side. The dam is still there, although its reservoir has completely filled with silt. Another similar dam was built at Harbaka, Syria in 132 AD, with a height 18m and length of 200m, but its reservoir is also now completely filled with silt, clearly illustrating the problem of sedimentation. Neither of these dams failed in the sense of structural collapse, but they failed to be able to store and supply water; hence, this excessive sedimentation can be regarded as a failure.

Lake Homs was formed in 284 AD by a 2km long low dam built across the river Orontes, in Syria. At its maximum section, the dam is 6m high, and consists of mass rubble masonry faced with blocks of basalt. It has been maintained and repaired throughout its life and is an example of how a dam, with correct maintenance, can remain useful for an exceedingly long period.

As the Roman Empire spread westwards, they built many dams in North Africa and Spain. In 193 BC, they gained control of Toledo and to provide a water supply for the city, they built the Alcantarilla dam, illustrated by Fig. 2.3, using the Nabataean design, but, unfortunately, without the gravel drain. It has an upstream wall of masonry about 1m thick, with a 0.6m thick core of rubble lime mortar supported on the downstream side by a second masonry wall also about 1m thick, in turn, supported by an earth fill shoulder, 14m wide at the crest with a 1 on 3 even downstream slope. This dam was 20m high and at least 550m long, and its remains can be seen even today, 12 miles south of the city of Toledo. Smith (1970) claims that it is the oldest dam in Spain and could be the earliest example of Roman dam construction. Its failure appears to have been caused by rapid drawdown, the central 220m length having fallen into a drawdown reservoir under the pressure from the downstream earthfill shoulder.

A dam of similar design was built 300 years later at Proserpina, 6 km north of Merida for water supply. It is 12m high with the masonry/lime mortar rubble/masonry sandwich 3 to 4 m thick at the crest, widening to 5m at the base, with the earth fill downstream shoulder having an outer slope of 1 to 5. It would appear that even after 300 years, they had managed to learn from the Alcantarilla failure and so supported the upstream masonry with nine heavy masonry buttresses, although there was no clear evidence of drainage. There are records of various repairs

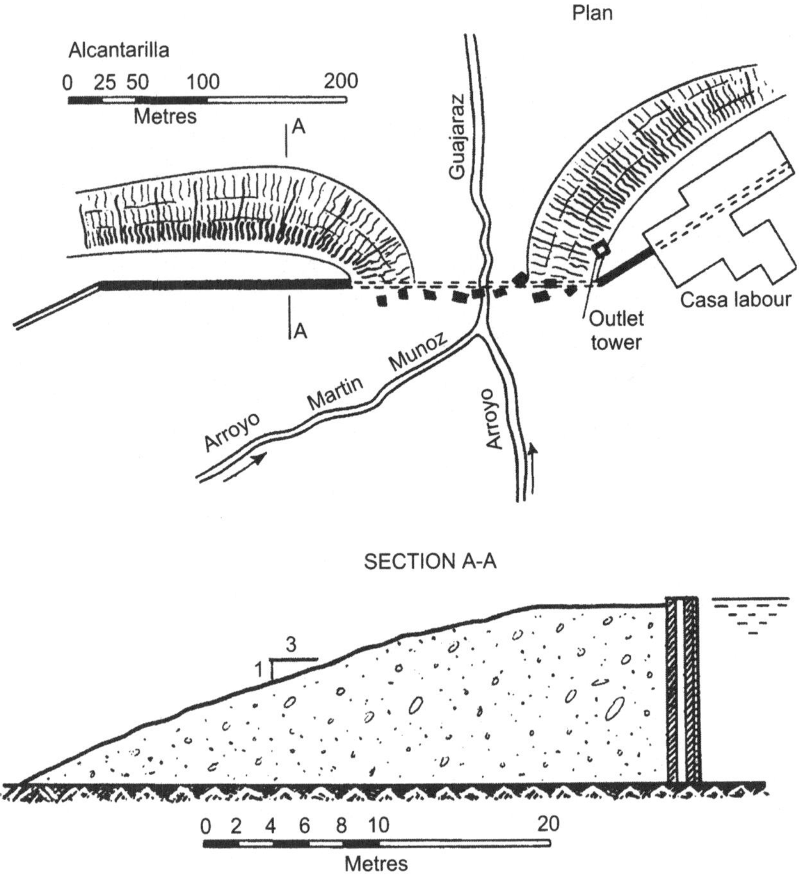

Fig. 2.3 Section of Alcantarilla dam, Spain

and modifications in 1617, 1689 and 1791, which have kept the dam fully operational to the present day.

Another dam was built for additional water supply for Medida on the Rio Albarregas, 16km northeast of the city. This Cornalvo dam is 20m high, 10m wide at the crest and about 200m in length. It is essentially an embankment dam with a core formed by two vertical masonry walls with cross walls forming square boxes filled with stones and clay. The earthfill shoulders have slopes of 1 on 3 downstream and 1 to 5 upstream, protected by rip-rap. This dam impounds a reservoir of over 10 million cubic meters, a capacity not exceeded in Spain until 1791. It was reconstructed in 1936, fitted with new valves and continues in service even today. Its design shows similarities with today's design for zoned embankment dams with central cores.

The Romans built many dams in what is now current Portugal, largely in areas of low rainfall where the dammed rivers were often dry and any supply of water depended on storage of the eighteen dams listed by Quintela et al., (1987) the highest was 5.2m, but the average height was only 2.7m. Almost half consisted of simple walls of masonry; one used a double wall with an earthfill core between, supported by buttresses on the downstream side; one supported by a downstream earthfill shoulder and the others as a simple wall supported by buttresses. None of these dams are in operation, the stream beds now passing either through or round them.

Cambodia (600 to 1200 AD)

The Khmer civilization was centered around the middle course of the Mekong River in the region known today as Kampuchea, just below the descent of the river from the uplands to the broad plains. This area became the junction of important overland trade routes and commerce and agriculture that flourished, requiring water control and irrigation schemes. Great catchment basins were formed by building embankments, often around entire cities and districts. The impounded water was released through a network of canals and channels to rice-growing land. An example is in the Surayararman II temple of Angkor Wat, that includes two major reservoirs. One of the reservoirs enclosed by a 10m high dam is 8 km long and 2km wide, and is till in use today. The main distribution canals extend about 60km and provide irrigation for about 13000 hectares.

India and Sri Lanka (500 to 1500 AD)

An irrigation scheme being built during the early 1980s on the Maduru Oya in Sri Lanka uncovered an earlier scheme dating from the fifth century. It was found that the position chosen for the new dam was very close to the site of an early embankment dam that had been 30m high. Details of this find have been given by Benson et al. (1983). Sections of the dam are shown in Fig. 2.4. Of particular interest were the twin outlet culverts, 66m long, passing through the dam at the right abutment so as to be founded on rock. They were constructed of brickwork lined by granite slabs forming passageways approximately 0.7m wide and 0.9m high. Carbon dating of a charcoal remains has indicated that the newest part of the structure dated from 460 AD. The cause of failure of the dam can only be speculation.

In India, there was a tradition of dam building that was once considered to be one of the seven meritorious acts which a man ought to

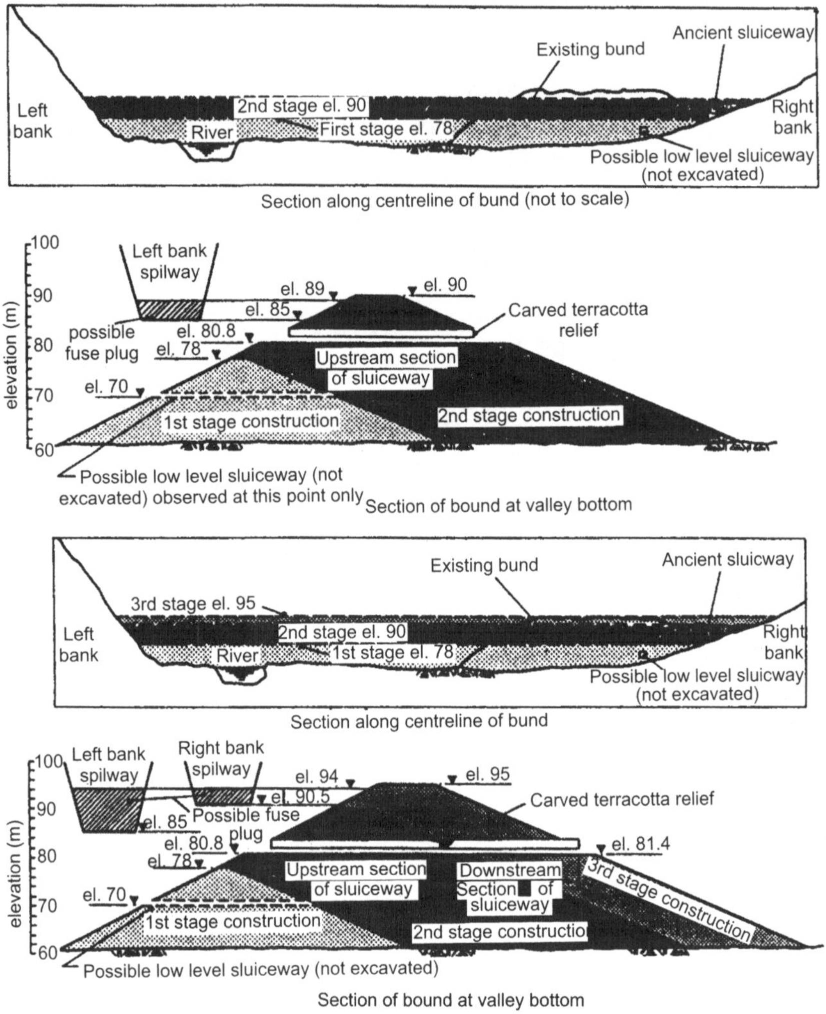

Fig. 2.4(a) The old bund and ancient sluiceway, showing (a) the second stage of construction, and (b) the third stage of construction

perform during his lifetime (Rao, 1951). Data given by the ICOLD World Register of Large Dams shows that by the time of the first millennium, India had five embankments dams with heights of 17 to 45m.

OLD DAMS [Essentially 1700 to 1900]

The accelerating increase of world population caused a growing demand for water storage and the need for dams. The true behaviour of dams was

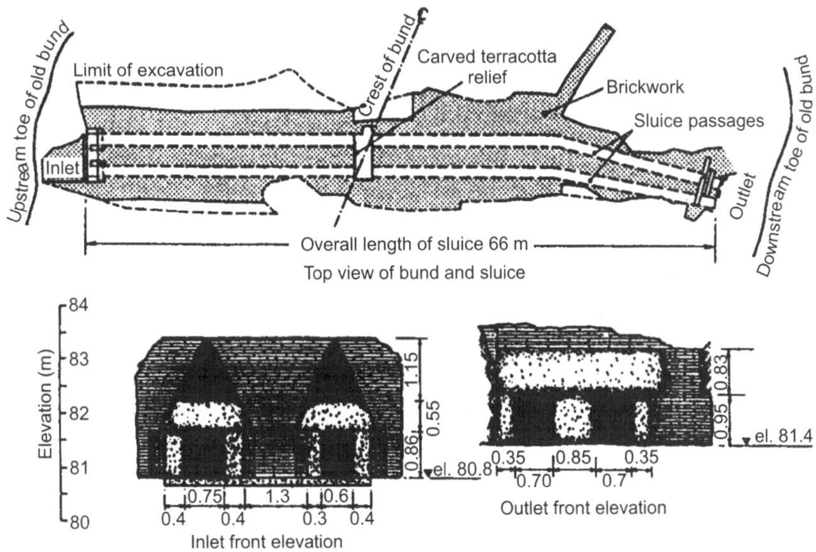

Fig. 2.4(b) Sluice openings and bund

not fully understood and design was based on precedent, for which the examples of failures and dam incidents have played a major role.

The Del Gosco dam in Spain followed the Roman pattern of two masonry walls sandwiching a core of clay and stones. It was supposed to have a height of 93m but during construction in 1788-89, it was overtopped at a height of 57m and the upstream wall overturned due to pressure from the water that had fallen into the core, combined with the pressure for the swelling clay.

The Puentes dam was another Spanish dam of the same period, but built of rubble masonry with ashlar facing to a height of 50m during 1785 to 1791. At its design height of 50m, it could retain a reservoir of about 40 $\times 10^6$ m^3. Its site had previously been considered suitable for a dam and, in 1647-1648, the foundation had been laid for a 26m high dam, but it was washed away in a flood. The Puentes Dam was in a narrow part of the valley of the river Guadelentin, with hard rock on both sides, but during excavation for the foundation, a considerable depth of alluvium consisting of earth and gravel was encountered in the bottom of the river. It was in general below the level of the water and with the equipment available at that time, it could have proved very difficult to excavate through the sandy gravely alluvium over the full 46m width for the base of the dam, capped by a timber platform to take the first courses of masonry, with a view to transmitting the weight of the dam down to bedrock. From the completion of the dam in 1791 until 1802, there was no more than 25m depth of water in the reservoir, and there were no signs of leakage. The water was allowed to rise, and early in 1802, it reached a depth of 47m when leakage did begin to appear. During the afternoon of 30 April 1802, a very red coloured water was seen bubbling up from the toe and immediately a messenger was sent to inform the director of works at the downstream town Lorca. After only about a half hour, explosions were heard and an increasingly massive flow of water, piles, beams and other woodwork came up above ground surface to be washed away in a great discharge accompanied by dislodged masonry, forming a hole so large that the reservoir was emptied in only an hour. The messenger had to escape up a nearby hill as the flood caught up with him, and it swept through the town causing an immense amount of damage. At the time of the final blowout, the water depth had been 33.4m. The hole was about 17m wide and 33m high, with the dam spanning over it so that most of the dam structure was undamaged, taken about 50 years after the failure. Interestingly, the foundation leakage had been only noticed 30 minutes before the failure at 15.00 hours on 30 April 1802 and the reservoir emptied in an hour, producing a flood wave that reached the city of Lorca 14km downstream in an hour, where it caused the death of 608 people and the destruction of more than 600 houses. The flood also caused devastation in the surrounding area where a further 328 were drowned.

The Bouzey dam on the river Avriere, France, was built in the form of a masonry dam, designed to resist the force from the retained reservoir water by its own weight: a type of dam referred to as a gravity dam. This dam was much wider at its base than at its crest, as shown by Fig. 2.5. The total weight of the masonry, acting through the centre of gravity of the dam, should be able to resist the overturning moment produced by the thrust from a full reservoir, as indicated by the force arrows given in Fig. 2.5. The foundation was a fissured red sandstone containing an almost horizontal bed of clay. Uplift pressure from the water was not a consideration at that time, and the width of the masonry was insufficient to prevent tensile forces developing in the upstream face. The dam was constructed to its design height of 30m during the period 1879 to 1882, and during its very first filling, tensile cracks developed near the base, allowing some seepage. When the water level in the reservoir had risen 12.3m, a central part of the dam, which was 135m long, moved downstream by 0.28m on the bed of clay. This caused further cracking and produced increased leakage, but the dam did not fail. It was four years before extensive repairs were carried out. The reservoir was emptied exposing a longitudinal crack near the base on the upstream side. This was covered by a block of masonry and sealed with puddle clay, while on the downstream side, the dam base was extended to double

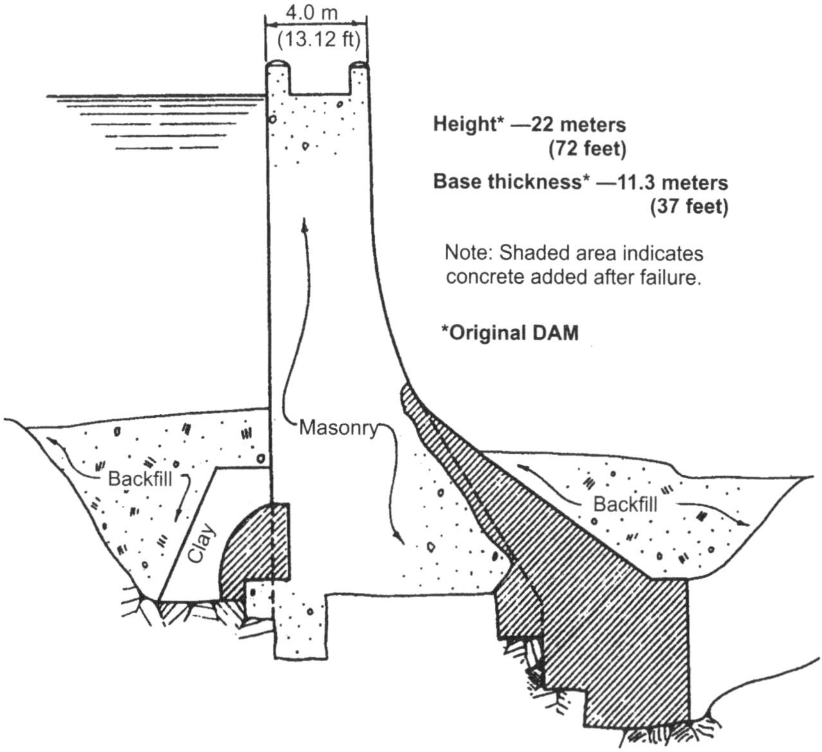

Fig. 2.5 Bouzey dam, France

its original width with a block of masonry forming a longitudinal buttress that was carried up to mid height of the dam. Immediately above this new buttress, the dam reverted to its original section, with no transition or special reinforcement. With the reservoir full, on 27 April 1895, a central part of the dam 171m long and 12m high suddenly overturned, releasing a catastrophic flood into the village of Bouzey and the Aviere valley, causing the loss of 150 lives. A lesson learnt from this failure was that allowance has to be made both for poor foundation conditions as also uplift any horizontal section through the dam at any level. It is now usual to reduce uplift pressures by drainage.

The South Fork dam, 100km east of Pittsburg, was built in three stages between 1838 and 1852 with an upstream shoulder that has been described as of puddle earth, supported by a downstream shoulder of loose rockfill, as indicated by Fig. 2.6. It was 21.9m high and suffered a blowout along its outlet in 1862. In 1880, it was bought and repaired by a fishing and hunting club and the spillway fitted with fish screens to

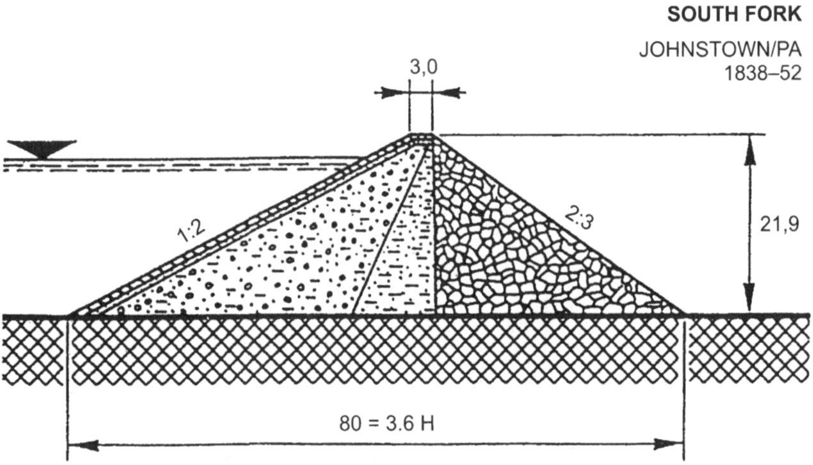

Fig. 2.6 South Fork dam, America

prevent loss of fish. This restriction to the spillway contributed to overtopping by a flood at the end of May 1889. After about 3 hours, the dam breached, releasing a flood that killed 2209 people. Prior to the Vijont disaster in 1889 this was the greatest number of people to be killed by a dam failure.

The Mill River dam in Massachusetts was a 13m high embankment dam, with a clay core. It was completed in 1865, but on very first filling, the downstream shoulder collapsed, due to what was thought to be water passing under the clay core and wetting the shoulder fill. No engineers had been employed for either its design or construction and this fact was used to suggest that both the design and the methods of construction had been faulty. This dam failed on 16 May 1874, the reservoir emptying into a narrow valley and causing the loss of 140 lives.

In Algeria, the Cran Cheurfae gravity dam was built with rubble masonry to its design height of 40m and completed in 1884. On first filling, it failed partially and was repaired. El Habra dam, also in Algeria, was of similar design with a gravity section constructed of rubble masonry between 1856 and 1871. It was 35m high and failed on its first filling, producing a flood that claimed 200 lives and destroyed several villages. It had been overtopped and a 90m long piece of the dam was eroded down to the foundations. It was repaired, but failed again in 1881, causing 400 deaths. It was repaired once more and raised to a height of 38m during 1883 to 1887 and served successfully for 40 years, then failed for the third time on 26 November, 1927 when it suffered a flood that overtopped it by 4m. Sig was another gravity masonry dam in Algeria, with a height of 21m. This was completed in 1845. When it overtopped in 1885 and failed, it was abandoned.

Saint Denis, 19m high, failed in 1885, causing 10 deaths. St. Lucien, 27m high, failed in 1862 by piping.

In England, the 20m high Bilberry embankment dam with central puddle clay core, built between 1839 and 1843, failed in 1852, causing the loss of 81 lives. The worst dam disaster was the failure of the 29m high Dale Dyke embankment dam on its very first filling, causing the death of 244 people in the Sheffield area. This failure triggered the UK Reservoirs Acts, which have turned out to be the best in the world. The story of that development will be narrated further in the book.

MODERN DAMS [Essentially after 1900]

Elwha River dam, Port Angeles, Washington, USA, completed in 1912, a concrete gravity dam in a narrow gorge, was 33.5 high with a crest length of only 10m. The bottom of the gorge was filled with alluvium that was excavated down to sound sand and gravel. Due to difficulties with dewatering, the bottom concrete had to be placed through water. The base of this so-called gravity structure was 30m wide and should have provided adequate stability. As the reservoir filled for the first time, on 9 October 1912, piping occurred through the foundation sand and gravel, washing a hole under the dam, similar to the failure of the Puentes dam, described earlier. The failure occurred in 1 ½ hours, leaving the dam spanning the gorge.

The concrete gravity Saint Francis dam in California, completed in 1926, was 62 ½m high. It failed when the reservoir was full to capacity at midnight on 12 March 1928, destroying 400 houses and causing 426 deaths. It had slid on its foundations and left sections of the dam standing erect, but downstream of their constructed positions. This failure was similar to that of the Tigra dam in India. This reservoir was also a gravity masonry dam 28m high, on a shale foundation. It failed in 1917 while being overtopped by only 0.15m, 30 minutes before the sliding failure. A sliding failure also destroyed the Austin dam in Texas, a gravity masonry dam 21m high, completed in 1893. It was overtopped and failed on 7 April 1900, sliding downstream on its shale foundation, the flood causing 8 deaths. The Austin dam was reconstructed in 1911, by replacing the damaged part with a hollow concrete section, but with a height of only 18m. It was damaged again by floods in 1915 and declared unsafe. Much later, in 1940, it was rebuilt to a height of 30m.

In the Czech Republic, the 14.5 m high Bila Desna embankment dam was constructed between 1912 and 1915. When the reservoir filled, there was no leakage from the dam itself, but clear water issued from a crack in the bottom of the outlet gallery. On 18 September 1916, when the reservoir water level was 1m below T.W.L., a leak of clear water was observed during the afternoon at 15.30. The leak was increasing and seen to be changing from clear to muddy, with more and more earth being carried by the water until at 17.00 hours, when the dam breached, releasing a flood that claimed 65 lives and destroyed a vast amount of property. The engineer responsible committed suicide. He had been told of the increasing leakage at 15.30 by the dam attendant and immediately ordered all outlets to be opened as well as sending a warning message to all the people living downstream. Unfortunately, the message was misunderstood, and instead of moving to higher ground, the people came down into the narrow valley to see the flood as a spectacle. They had no chance of escape when the released flood swept through the narrow valley. A section of the dam is presented in Fig. 2.7, showing the outlet culvert along with the erosion and breaching occurred. Culverts taken through dams are a severe danger. They have been several examples of breaches occurring along the line of such culverts. The modern dam regulations in Italy forbid the use of culverts or pipes passing through embankment dams from upstream to downstream.

In Spain, the 34m high concrete buttress Vega de Tera dam was completed in 1956, 5km upstream from Ribadelago. Three years later, during the night of 8/9 January 1959, there was a collapse of the buttresses, releasing a flood that claimed 144 lives.

There are three dams on the Rio Pardo in Brazil. That furthest upstream is the 60m high Caconde dam with a bellmouth spillway built during 1959 to 1966 that impounds $550 \times 10^6 m^3$ in a reservoir with an area of $31km^3$ and produces a power output of 80.4 MW. Further downstream is the 63m high Euclides da Cunha with a gated spillway of 2,040 m^3/sec. capacity, constructed during 1956-60, impounding $12 \times 10^6 m^3$ and producing 94.8 MW. The third dam is the 41m high Limoeiro with a gated spillway of 1,800 m^3/sec. capacity, and the first to be built, during 1953-58 that impounds $25 \times 10^6 m^3$, producing 28 MW. All of these dams are embankment dams with the Brazilian section. During floods in 1966, a flow of 800 m^3/sec. discharged safely though the gated spillway of Euclides da Cunha, inundated the ground level power station of the Limoeiro dam and drowned a downstream chicken farm. This caused a restriction to be imposed on the permitted gate openings of the spillways of both the Euclides da Cunha and the Limoeiro dams. Eleven years later, during the period 12-18 January 1977, there was very heavy rainfall and then, on 19 January, 230 mm fell in 24 hours. When the resulting flood of 2,000 m^3/sec. reached Euclides da Cunha, the reservoir supervisor tried to obtain permission to exceed the restricted gate opening allowance.

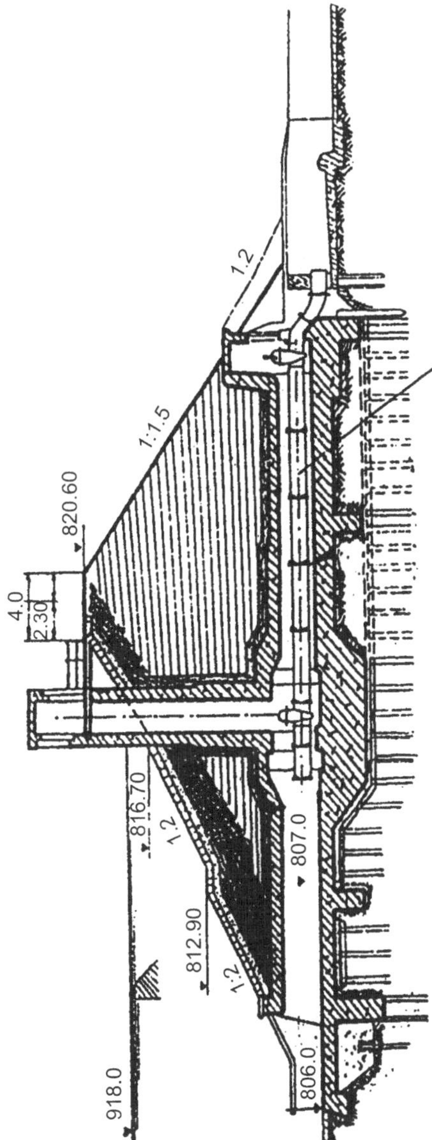

Fig. 2.7 Bílá Desuá—Czech Republic

Because of the storm, telephone lines were down and he set off to try to find a working telephone. Meanwhile, the reservoir level began to rise rapidly until the dam itself began to be overtopped. Settlement since construction had caused the central part of the crest to be 0.3m low, so the overtopping began there. The reservoir level continued to increase until there was 1.3m head of water passing over a 450m length of crest which continued for 7 hours before the dam breached. The released water overtopped and failed the Limoeiro dam and the even greater volume of released water travelled down the Rio Pardo, joined the larger Rio Grand and was absorbed by the $6{,}150 \times 10^6 m^3$ reservoir retained by the 96m high concrete Marimbondo dam. The required larger discharge from its spillway that had a $21{,}400 m^3/sec.$ capacity, and was in turn, absorbed by the $11{,}000 \times 10^6 m^3$ reservoir retained by the 63m high Agua Vermelha, the next dam downstream on the Rio Grand. Both the Euclides da Cunha and Limoeiro dams were rebuilt during 1977 to 1980 with the spillway capacity increased to 3,100 m^3/sec. for the first and to 3,000 m^3 sec. for the second dam, respectively.

The 71m-high Tous entire embankment dam was severely overtopped and failed on 20 October 1982. The entire area was suffering from severe flooding, and the spillway gates could not be opened. Overtopping began at 17.00 hours, and by 19.15 hours, the depth of water passing over the dam was 1.1m, causing erosion of the downstream shoulder until failure occurred. Downstream, 8 persons were killed, although a great deal of damage was done by the flooding prior to the arrival of the flood wave from the failed dam.

In Italy, the 15m high multiple arch concrete Rutte dam was built during 1951-1952 at a very advantageous site for hydro-power, at the edge of the Gereuth plateau where it falls away steeply to the Gaillitz valley near Tarvis. The power plant at the foot of the precipice required a large side cutting into the precipice, and soon after impoundment began, seepages were seen at the dam toe and on the face of the precipice. A crack in the floor of the reservoir was sealed and some grouting carried out, which stopped the seepages. By the end of 1965, silt carried into the reservoir was 9m deep and it was decided to clean the reservoir, which was emptied. The sealing of the bottom crack was renewed and impoundment began again on 5 November 1965. At the position of the crack, the dam toe was founded on crumbly dolomite, and two days after the start of impounding, it was undermined, forming a 2m high and 12m long hole. Two buttresses collapsed, allowing their arches to fail and the reservoir emptied in 65 minutes. The maximum discharge was estimated to have been 160 m^3/sec, but fortunately, warnings to the Austrian Departments downstream enabled the flood to be accommodated in their hydro-power reservoirs.

MASONRY DAMS

In early days of dam building in India, masonry dams constituted a majority of the dams and were constructed with masonry retaining wall on the water side and earth backing on the downstream. This practice was mostly confined to the erstwhile Hyderabad State, in southern India, as surveyed by Murti (1978, 1979), where even now such dams are in service.

Kaddam Dam

The Kaddam dam, built in 1957, was perhaps the last dam with masonry facing and earth back filling on the downstream. The masonry wall, about 40m high, was very slender, 0.46 H at the base and without any drainage arrangement or filters in the earthen backing. In 1959, the reservoir water rose above the MWL and the dam experienced overtopping. This resulted in washing away of the earth backing and collapse of the masonry portion. Reconstructed with an enlarged spillway, the earth backing was maintained and the dam is still standing as the last example of this type of masonry dam. Perhaps this method worked out to be less expensive in those days, saving the upstream slope of an earth dam, substituted by a masonry wall of 2/3 size. The practice of earth backing as a part of initial design is no more in existence, but it is in vogue for the restoration of old masonry dams, where uplift pressure or earthquake forces were not taken into consideration in the initial design for design purposes.

Chikkahole Dam

Situated in the state of Karnataka, in southern India, the Chikkahole dam is a hand laid stone masonry dam, about 670m long and the maximum height from the deepest foundation is 30m. The dam is constructed in lime surkhi mortar. The foundation geology is weathered granitic and biotic gneisses, which required consolidation grouting before the construction of the dam was taken up. A mishap that occurred during grouting—indicated by a loud sound—caused a vertical crack in stone masonry, followed by certain minor cracks near the right flank. These cracks were attributed to the grouting pressure. Intensive grouting was then taken up to fill up the cracks that later appeared.

About six years after construction, on 8 December 1972, the dam breached suddenly on the right flank, without any sign of distress. Investigations were conducted to examine the reasons of failure. One of the probable reasons was that the breach had occurred where the dam had already developed a horizontal crack on the upstream face due to uplift pressure caused by grouting. Although additional grouting was done, it did not add to the bond strength at the joints because patches of grout material, with air voids and free and smooth top surface, were noticed at various portions during post-failure inspection. Perhaps the tensile cracks progressively extended into the body of the dam and along the face of the masonry, prior to the breach. As the cracks progressed deeper into the dam, the uplift within increased, adding to the compressive stress on the downstream face, which resulted in development of tensile stresses in perpendicular direction. This caused a reduction in the section of the dam and ultimately resulted in tensile and overturning failure. The sudden failure is attributed to initial cracking of the masonry. According to Murthy, Mane and Pant (1979), this failure phenomenon was proved by model studies also.

As far as the design is concerned, the density of the masonry was assumed as 2.49 tonne per cubic metre and full uplift acting on 2/3 the area at any level. Based on these assumptions, tensile strength is not expected in the section. The maximum compressive strength in the masonry at full reservoir level was assumed as 9.5 kg/cm^2 at the downstream face and 0.6 kg/cm^2 at the upstream heel. In reality, the density of masonry of the dam was of the order of 2.2 t/cm^3. A highly permeable masonry, without any drainage arrangement, can develop full uplift on the entire area. Calculations showed that under full reservoir conditions, with density of masonry ranging from 2.2 to 2.4 t/cm^3 and with full uplift at the upstream face reducing to zero at the toe, there is the possibility of a high tension zone developing around the upstream face of the dam. With reservoir water reaching the top of the dam, the situation must have been worse at failure.

Lime surkhi (lime + burnt brick powder) mortar used for the masonry of the Chikkahole dam was 1:4 and the compressive strength specified at 12 weeks age was 28.4 kg/cm^2. The strength of the mortar used varied from 10.5 kg/cm^2 to 51 kg/cm^2 giving an average strength of 29.4 kg/cm^2. The split tensile strength varied from 1.5 kg/cm^2 to 3.9 kg/cm^2, giving an average of 2.6 kg/cm^2. At Chikkahole dam, the major cause of reduced tensile strength was the cracks that had developed in the stone masonry, perhaps due to over pressure grouting. High compression on the downstream face masonry separated it from the hearting masonry by the tensile failure of the bond between the face stone and the hearting. The masonry was taken as impermeable, ignoring the wetness noticed on the downstream face. It is not the quantity of leakage in a masonry dam that counts, but the extent of uplift pressure developed that causes distress. The red surkhi mortar used in the dam also had the tensile strength of a lower order than the one specified.

The controversial reasons, the strength of mortar and the grouting, which was not a unique practice adopted for strengthening the foundations, were perhaps confined to the failed portion. It is interesting to note that the portion of the dam which did not fail was proposed to be strengthened by providing earth backing of about 9.0 m top width, as a part of restoration of the dam.

Bhandardara Dam

One of the highest and oldest masonry dam in Maharashtra state in western India, the Bhandardara dam is a gravity structure built in rubble masonry, using hydraulic lime mortar containing coarse and crushed sand. The dam has a Wegman's profile without any drainage arrangements. Constructed in 1926, with a height of 82m above the deepest foundation level, the reservoir was storing water every year to the designed reservoir level. The dam had been exhibiting extensive seepage and wetness over the entire downstream face. But on 16 September 1969, from a 15cm drain outlet, water started jetting out by about 6m on the downstream. Till that day, water was trickling from the same pipe and did not attract attention. In the backfill, large areas with signs of boiling and water springs were also noticed and total discharge of the order of $0.87 m^3/s$ was recorded. Vertical cracks going upto 0.6 to 7.0m were also noticed above the foundations. When the backfill was removed at the toe, a sheet of water gushed out from joints between the stone masonry and foundations. Investigations were thus taken up to locate the source of leakage and development of cracks, if any.

Drilling commenced to locate the crack and its pattern. At a depth of 44m, close to the upstream face, there was a total loss of water. Further from the flow of cement, grout free passage of water was established at this level from the upstream and at rock-masonry contact at the toe. This was further confirmed by sodium chloride tests. The crack path—in a similar fashion—was established by resorting to drilling from the downstream face. By reducing the water level of the reservoir, the upstream tensile crack became apparent. On the downstream side, the crack could be observed through an adit excavated on the downstream face.

Tests on masonry cores from the Bhandardara dam gave an average compressive strength of the order of $78 kg/cm^2$ and modulus of elasticity as $0.85 \times 10^5 kg/cm^2$. The average density of masonry was $2.4 t/m^3$, which was found to be equal to that of mortar (lime: sand being 1:2). Compressive strength of mortar varied from $80 kg/cm^2$ to $120 kg/cm^2$. The tensile strength was about 1/8 of the compressive strength. The cracks on the upstream face were wider in the centre and indicated signs of movement towards downstream by 3mm to 6mm. No vertical cracks were observed on the upstream face. The central tensile crack, acting like an arch had presumably transferred a part of water load to the abutments, having sustained this without causing any distress. The hydrological safety consideration, it is reported while reviewing the safety of the old dams—had necessitated additional spillway capacity. But before a decision was taken, a seismic event occurred in 1967, 'followed by some unusual manifestation in the body of the dam. However, by providing drainage holes on the downstream and grouting of the upstream, there was some time to take up adequate measures to strengthen the dam.

Although the dam did not fail in the literary sense, yet it could have but for the timely indication of heavy leakage through a pipe. The investigation in the right direction, both on the field and by model studies, leading to detection of crack and lowering of reservoir water level, saved it from collapse. The dam had been leaking throughout its 40 years life; the slow progress of the tensile crack took time to increase uplift pressure. It is also felt that in plan, the arching upstream of the dam also helped to sustain certain undesirable pressures for a longer time. The masonry and mortar did not indicate any sign of deterioration, although wet patches and sweating were noticed right from the beginning.

Machhu-II Dam

Machhu basin in Gujarat State, in western India has two dams; namely Machhu-I and Machhu-II. Machhu-I, constructed in 1959, is a masonry dam of 28m height, whereas Machhu-II is a composite dam, constructed in 1972 with the 206 m long masonry spillway in the center and a height above the deepest foundation of about 25m (revised to 27.5m). On the right and left of the spillway were two earth dams, 1524m, and 2327m long respectively. The spillway was designed for $6182 m^3/s$ discharge capacity with its crest level at R.L. 51.20m and had 18 numbers of 9.14×6.1m size radial gates. The FRL and MWL were at R.L. 57.3m and 57.60m, respectively, with free board as 2.74 (revised to 6.40) at FRL and 2.45m (revised to 4.5) at MWL.

On 11 and 12 August 1979, the Machhu basin experienced the major thrust of a storm which had hit the area. The nearest raingauge recorded rainfall of the order 518.8 mm in about 20 to 22 hours between 10 August night and 11 August evening. It was an unprecedented flood and earthen embankment on the two flanks, about 700m on the right side and 1070 m on the left, were washed away due to overtopping on 11 August at about 14.30 hrs. This overtopping indicated insufficiency of spillway for the flood of the order of $14000 m^3/sec$. An 8m high flood wave brought with it several meters of mud and debris, causing unparalleled damage in the nearest Morvi town in Gujarat which was famous for its parks and avenues. The estimated loss of life was of the order of 2000 besides property loss of an order of 1160 million rupees.

The revised design flood has been taken as 26420 cumecs and additional twenty radial gates of size 12.5×8.23m, thus, became necessary. The earth dams on both the flanks have been designed reach wise. Besides the new earth dams, the existing earth dams were raised to adopt to the revised standards by shifting the old center line towards downstream wherever necessary.

CONCRETE DAMS

Hirakud Dam

The Hirakud dam in Orrisa, India, is a composite dam. It consists of earth, concrete and masonry dams across the river Mahanadi with the main dam of 4800m length and concrete and masonry dams about 1149m long gives the plan, sections and other particulars of the dam and appurtenant structures. The dam was constructed during 1954-64 and in the following year distress was observed, leading to stoppage of further constructions and a search for the causes and ensuring remedial measures.

- Extensive horizontal cracks, about 12mm wide, were observed on the upstream face of the dam. Horizontal cracks on the face of the gallery gate shaft and sluice barrel in the right side spillway blocks were also observed. The foundation gallery was also affected to some extent. The maximum width of cracks at the surface was of the order of 6 to 9mm, extending to a depth of 2m.
- Deflection in Adit gallery was noted.
- Manhole frames opening on the roadway got buckled.
- The bolts, fixed to the guide rails of the sluice gates, were snapped or sheared.
- The steel frame fixed above the collapsible gate, as entry to the operation gallery, was buckled.
- On the downstream face of the right spillway, horizontal cracks at various levels and vertical cracks in two blocks were observed and their widths appeared to be increasing slowly.

The investigations taken up by the National Council for Cement and Building Materials (NCCBM) found the concrete to be good quality and was well compacted. However, the concrete of the right spillway, as revealed by tests on concrete cores, had suffered an 'alkali-silica' reaction. This reaction was very obvious by the presence of white, translucent to opaque reaction products on the surface of the aggregates, in the cracks and hollows of the concrete. Petrographic analysis and microscopic scanning further confirmed the reaction. It is now well established that alkali component in cement and certain active minerals, in aggregate, do cause such reactions, and result in deterioration and endangering the durability of a structure.

Remedial measures included sealing of the cracks on the upstream face and grouting of the dam using a suitable chemical grout. Chipping, bulging and deflections to the possible extent was also suggested.

Rihand Dam in north India, about 91.5m high and 934.5m long, was constructed during the period 1954-1962. In 1972, certain distress signs were observed, which were spread out a different locations and in a different pattern. Investigation by NCBM revealed that the concrete used was of good quality and was not attacked by sulphates. The cracks and deterioration in concrete was found to be the result of alkali-silica reaction. Aggregates quarried for concrete used in the body of the dam contained strained quartz and alkali felspar. The presence of gel type reaction products inside the concrete and similar other manifestation on the aggregates, and petrographic examination and scanning by electron microscope also confirmed the alkali-silica reaction.

Talakalale Dam

Talakalale dam, a masonry gravity dam constructed during 1959-63, has a maximum height of 62.5m above the deepest foundation level and 354m in length. The dam forms the balancing reservoir of the Sharavathi Hydro Electric Project and is thus supposed to maintain water at full reservoir level for generation of maximum power. The mortar used in the masonry is of two types. The bottom portion of the dam, about 50m in depth, has been built with stone masonry using cement surkhi mortar (cement+ sand+burnt brick powder) and above it, the mortar used is lime-surkhi mortar.

The masonry on both the upstream and downstream faces are of standard size, with random rubble masonry in the hearting of the dam. Vertical porous pipes, at intervals of about 6m, are embedded in the masonry and connected to 100m long drainage gallery to drain out the leakage or seepage water by gravity flow from body of the dam. There are no sluices in the body of the dam and the central spillway was constructed as originally designed.

The first monsoon filling of the reservoir in 1963 was restricted to RL 479.00, i.e. to a depth of 21.65m against the full reservoir level 516.65m. Even at the 1/3 filling, heavy leakage was observed on the downstream face of the dam throughout the length. After building gates in 1963-64, water was stored to its full reservoir level. Besides the increase in seepage/leakage on the downstream face, water was sprouting in many more places.

Grouting in selected portions of the dam in 1964-65 on the downstream face did prove effective to control the spouts but the seepage/leakage did not decrease. The entire length of the dam was then grouted in 1965-67, which reduced the seepage from 3.91 cusecs to 0.23 cusecs. But the increasing trend in seepage/leakage continued from 1970 and in 1974, it measured as 0.635 cusecs.

In 1974, the upstream face was examined by a close circuit TV, which revealed several open joints between the depth of 12-25 m from FRL. Out of 88 such locations, suction of water by mortar was not felt in 20 locations. Potassium permanganate tests revealed, from the coloured water that emerged on the downstream face, that the leakage of water was interlinked. The water from the body of the dam was diverted to drainage gallery by drilling three experimental bore holes. This brought the water gushing into the gallery, which were immediately plugged. Grouting from top of the dam was found ineffective, as the grout material was not retained in the masonry. A relatively cheaper solution—to stop leaching of the mortar—was found in earth backfilling on the downstream of the dam when compared to additional concrete on the downstream face or pre-stressing dam by cables.

To keep the seepage within limits, various measures were examined and underwater pointing and subsequently grouting of the dam was favoured for various reasons. The underwater pointing alone brought down the seepage to 0.25 cusecs. The dam was extensively instrumented by installing settlement gauges, earth pressure cells and standpipe piezometers to monitor the performance.

There were certain doubts expressed for the restoration design regarding stabilizing force to be assumed—whether active earth pressure at rest—particularly as the dam will not be subjected to a draw down conditions. Based on experience and studies made elsewhere, it was concluded that compacted earth fill exert earth pressure at rest. The top of the earth backing was kept 9.14 m at FRL, which could have been lowered—as suggested—by 3m in view of no draw down and no yielding of masonry dam can be expected. Due to the earthen backing on the downstream, the surplussing arrangements were also shifted in 1976-79.

EMBANKMENT DAMS

Tigra dam

Constructed across the River Sank, a tributary to river Chambal in Central India, Tigra dam is a 24 m-high masonry dam, about 341m in length, of which 193m is earthen. The dam was constructed in the period 1909-1917 in rubble masonry in lime mortar with cement joints on the upstream face to serve as a source of water supply to Gwalior city. The dam had an uncontrolled spillway, 305m in length and crest at an elevation of 225.6 m. The elevation of the top of the dam was 226.2m with parapet top at an elevation of 226.9m. Initially, the dam was designed for maximum flood 1455 cumecs. Maximum and full reservoir level were 225.5m.

Apparently, the dam foundations consisted of stratified sandstone of quartzite nature and the dam was founded at a depth of 0.6m. The weaker zones in the foundations were filled up with concrete to a greater depth.

On 4 August 1917, the dam failed, resulting in washing away of some masonry blocks along the river course during its first filling. It is reported that the body of the dam had developed cracks. On the day of the failure, the flow over the dam, from one abutment to the other, was of an order of 8500 cubic meters per sec. and depth 0.3 meters over the top of the dam. One of the views was, that under this water pressure, the dam had virtually slided over the base, causing the breach of the greater portion of the spillway. The two undisturbed blocks of stone masonry, which slided on to the downstream of the dam line, stand even today as a monument of the good quality of the masonry work.

The other view was that the failure of the dam was due to tension associated with hydrostatic uplift pressure and the scouring of the backfill due to the prevailing overflow conditions.

The Tigra dam was later reconstructed between the period 1919 and 1929. Two years after re-commissioning of the dam, cracks reappeared in the non-overflow section, resulting in seepage through the body of the dam. The geologists ruled out the possibility of seepage and any foundation movement due to sandstone formations. As remedial measures, the dam was redesigned for maximum flood of 5500 cumecs, and recommen-dations by expert panel included guniting of the upstream face of the masonry and increasing the surplussing capacity.

Kodganar dam

Kodganar dam was constructed across the River Kodganar in Cauvery basin in Tamil Nadu, south India. The 2662 m-long and about 16 m-high earth dam included a 231m-long masonry spillway with 5 vents of the size of 12 × 3.05 m. The design flood was of the order of 2,266 cum. per sec. The catchment area of the dam is of the order of 1670 sq km and the maximum water level was fixed as 201.75m and free board as 2.5 m. The dam was constructed in 1974-1977 and failed on 12 November 1977, during its very first filling.

Dam failure

In November 1977, the catchment of the dam experienced unprecedented rains accompanied by cyclonic conditions and cloudburst. About 135 small reservoirs on the upstream of the Kodganar dam were already full in September and October 1977 and the additional floods of November caused a breach of these dams. On 12 November 1977, at about 6.00 p.m., large floods reached the Kodganar reservoir, resulting in the threats to the safety of the reservoir dam. To allow the floods, when the spillway shutters were being lifted, there was electrical power failure. Consequently, the stand by diesel powered generators were put on, which did not prove very useful as these generators also failed. Subsequently, the last resort was manual operation. As a net result of all these efforts, by about 08.30 p.m., all the shutters were fully raised and the spillway was heavily discharging the floods. But the reservoir continued to receive water, causing a rise in the water level and ultimately at about 10 p.m., the dam was overtopped. It was reported that the entire length of the dam overflowed for about an hour. During this period, jumps were noticed on the downstream slope, resulting in heavy erosion. At about 11 p.m., at the ultimate stage, the dam breached in four reaches, from RD 185 to 205m, RD 300 to 450m, RD 1280 to 1480m and RD 2085 to 2130m. Other portions were also badly damaged with the downstream slope badly eroded and a gully formed. However, the masonry structure, the breast wall and the regulator in the riverbed remained undisturbed. The three counter-weight beams, of vents 2, 3 and 4 all washed away. The two canal sluices functioned normally and remained intact.

The depth of cutoff below the river bed was about 4.4 m. There was no failure of foundations observed.

Causes of failure

There were cyclonic conditions and unprecedented heavy rains in the catchment area and the river carried heavy discharges. The more significant added factor was the breach of small reservoirs on the upstream, which increased the flood discharges.

The Enquiry Committee later examined the causes of failure and recommended a revision of design flood waters to 8500 cumecs, against the original design flood of 2,266 cumecs. This meant an addition of 10 vents of size 12.2 m × 8.5 m at RD 1280 m to 1480 m.

Chandora Dam

The Chandora earth dam, about 2043 m long and 17 m high, on the two flanks, had a 67 m-long gated spillway across the River Tapti in Madhya Pradesh in central India. The dam was constructed during the years 1979 and 1986. The dam breached on 30 July 1991 after unprecedented rain in the catchment area, measuring about 71 sq.km. The reason given for the failure was that the central masonry ogee spillway had 8 numbers of radial gates of size 5.50 × 3.50m. The designed spillway capacity was 681 cumecs and gross storage of the order of 18.20m. cum. The dam was provided with a head sluice, RCC barrel of size 0.91 × 1.52 m, with a discharging capacity of 2.00 cumecs.

Failure of dam

As stated above, the Chandora dam failed in July 1991 after unprecedented rains—about 450mm in 24 hours—which caused overtopping of the dam. The upstream slope was damaged from RD 570 to 810m. The breach occurred, covering full height of the dam and foundations, from 810 to 960m. On the downstream side, overtopping had caused deep cuts from RD 960 to 1400 m. Even after restoration, cracks were noticed in May/June 1992 on upstream slope between RD 615 and 780m, which were then repaired before the reservoir received water. During March/April 1993, slips were noticed on the upstream slopes when the water in the reservoir had depleted. The slope was thus provided with rip-rap on the upstream slope between RD 615 and 810m. During repair work, cracks reappeared in March/April 1994 and the solution was to remove the rip-rap upto RL 685 to provide toe-loading berm.

Causes of failure

The unprecedented rains which had caused overtopping of the dam may be one of the reasons of failure of the dam. But the lips in the slopes and cracks on the body of dam may be due to settlement of embankment or the foundations. As per design, the dam was supposed to rest on hard rock, but during investigations, the foundations in failure reach RD 625 m to 660 m of the dam had compressible stratum of plastic clay and silt. Another view was that the dam was not designed for a sudden draw down and the plastic soils used were not provided with upstream filters. The cracks of 1991-92 were attributed to construction pore pressures caused due to a higher rate of construction, which in earlier stages were controlled due to slow construction.

Shiposhi Earth Dam

The Shiposhi earth dam in Ratnagiri District in Maharashtra, India, is a minor earth dam with a storage capacity of 2.6m.m^3 and maximum height of 24m. The gorge filling of the dam was commenced in March 1978 and completed in May 1978. In the following first rainy season, the reservoir was full, the water was standing at full supply level and there were no obvious distress conditions in the dam. An outlet conduit was provided in the body of the dam with stanching rings around the conduit pipe.

Failure

In the year 1985-86, for the first time, leakage of water was observed near the downstream head wall. With the varying storage level in the reservoir, the discharge varied from 0.04 to 0.05 cum. The various remedial measures taken up could not control the leakage, and the dam breached at the outlet. A cross-section of the earth dam shows the location of the out-let.

Causes of failure

- A review of the design showed the probability of the formation of a hydraulic jump within the outlet conduit during operation of the gate. The vibrations might have opened the joints of the conduit, resulting in leakage of water. The slope of the outlet pipe as adopted was flatter than the critical slope.
- The scheme classified as a part of the minor irrigation. It is also suspected that proper quality control might not have been observed in the compaction of the soils around the conduit pipe. The loose or poorly compacted soils must have migrated once the joints were opened and water started to leak.

ARCH DAMS

The weight and hence volume of material needed to construct a gravity dam is considerable because the weight of the dam is required to counter the overturning movement produced by the trust from the retained reservoir water. This weight, volume and associated cost can be reduced by providing buttresses to support a thinner and lighter dam. But, by using the resisting thrust from the valley sides in order to counter the reservoir water thrust, a vastly thinner dam can be made in the form of an arch. It behaves like an arch bridge on its side. The thrust from the water transforms into thrust from the ends of the arch on the valley sides, and provided that they are strong enough, a remarkably thin arch can be used. Unlike a bridge, an arch dam is not a pure arch because it is supported along the lower edge where its foundation goes into the ground. This makes the stress analysis quite complex and, because of the variations of valley shape, each dam is an individual and must be designed separately. While the early arch dams were considered as arches in one plane, the advantage of a three-dimensional approach was soon recognised. If the dam was a rubber sheet able to take only tensile stresses, it would bulge under the pressure from the retained water, rather like a hot water bottle. If that same shape is constructed in thin concrete and turned round to face the water, with the bulge towards upstream, then the concrete should be in pure compression, a force which the concrete can take. Such a design can produce remarkably thin dams.

One of the leaders in arch dam design was Dr André Coyne, founder of the well-known firm of consultants, Coyne et Bellier in Paris. Prior to his design of Malpasset dam, Coyne had been responsible for the design of more than 30 successful arch dams. The Malpasset dam on the River Reyran upstream of the town of Frejus, was 60 m high with a crest circular in plan with a radius of 105m, sustaining an angle of 133 from abutment to abutment. It was 1.5m thick at the crest and 6.78m at the base and was designed as a double curvature arch dam. On the left side abutment, the rock was excavated to a depth of 6.5m in order to provide a concrete thrust block. Impoundment began in 1954 and the reservoir was slow to fill. Movements of the dam were carefully measured by geodetic survey on to 28 targets on the downstream face. In general, the movements were in good agreement with prediction. At the major section, the base had moved out only 10mm more than predicted and the worst discrepancy was 15mm.

In July 1959, the water level was still 6.3m below the spillway ogee, but towards the end of the year, in late November, there was a period of heavy rain and the reservoir began to fill more rapidly. A new highway bridge was under construction over the river downstream and on 2 December, at 18.00 hours, the bottom outlet was opened to ease the rate of rise, so that the overflow from the spillway, when it came, would be less. Nothing abnormal was seen and even the caretaker who was painting on the crest observed nothing untoward. Yet that same evening, the dam failed in a single sudden movement. It was dark and there were no direct witnesses closer than 1½ km downstream near the bridge site. They felt the ground tremble, heard a brief rumble like the bellowing of an animal, then there was a blast of air followed by a wave of water that carried away the temporary bridge supports and, finally, a tremendous wall of water which could not be seen properly because by then the bridge lights had been extinguished. Pierre Londe (1987) summarized the main characteristics of the failure as: (1) a sudden explosion; (2) practically the whole of the dam was destroyed; and (3) nothing abnormal had been detected on the dam just a few hours before the failure. The town of Frejus was severely damaged and about 400 people lost their lives.

This was the first failure of an arch dam. It could be said that the dam did not fail: the failure was of the support from the left abutment which gave way. There had been an undetected tectonic fault with its strike parallel to the chord of the arch and dipping 45 and, due to the fissuring of the rock, it could be subjected to high pore pressures when the reservoir was full. The failure was a terrible blow to Dr Coyne. He was due to give lectures on arch dam design both in Paris and to the

Institution of Civil Engineers in London not long afterwards. Needless today, he did not give the lectures and, in fact, died only a few months afterwards, on 21 July 1960. André Coyne was born in 1891, served in the French Air Force during the 1914-18 war, then was in the Public Works Department in the 1920s with the port of Brest when he became involved with dams. A newly formed section of the Department of Service d' Aménagement de River Dordogne called Service Technique de Grand Barrage was set up in 1935. Six years later, coyne became Inspector-General de Ponts et Chaussées where he remained until his retirement from government service at the end of the 1939-45 war, when he set up in private practice. He was taken ill while on a visit to the High Aswan Dam in Egypt in June 1960 when he was rushed home for a serious operation.

BALDWIN HILLS DAM

The reservoir was on the southwest edge of the Los Angeles Basin, an area that had suffered severe tectonic deformations during the Late Pleistocene period, and has been subject to earthquakes; the most recent in 1933 when an M 6.3 event occurred. The 71 m high homogeneous embankment dam shown by Fig. 2.8, was completed in 1951. Late in the morning of 14th December 1963 the caretaker at the dam found an unusual flow of water from the drains under the reservoir lining, and a programme of reservoir lowering began. But within two hour, in daylight, muddy water was seen leaking away downstream. As the reservoir level feel, a rupture could be seen in the reservoir lining in line with the muddy discharge, and soon afterwards a breach formed through the dam that cut down so that the whole reservior was emptied. The reservoir was built in an unpopulated area, but the good access road built for it attracted housing developers and houses had been built just downstream of the dam. Because the failure occured during daylight, ample warning could be given and the only loss of life occured to someone who went back to save their canary.

DAMS ON WEAK LAYERS

Thorpe Marsh Flood Bank

An early example of analysis using effective stresses was the failure of a small flood bank in 1948. The 5.5 m bank, shown by Fig. 2.9, was built of a brown clay with an undrained strength c_u = 29 kN/m^2 on a foundation of a softer blue clay containing a peat layer with c_u = 13 kN/m^2 on the left side of a new channel to be excavated in order to remove a meander from the River Don. It had been built from borrowed fill; the

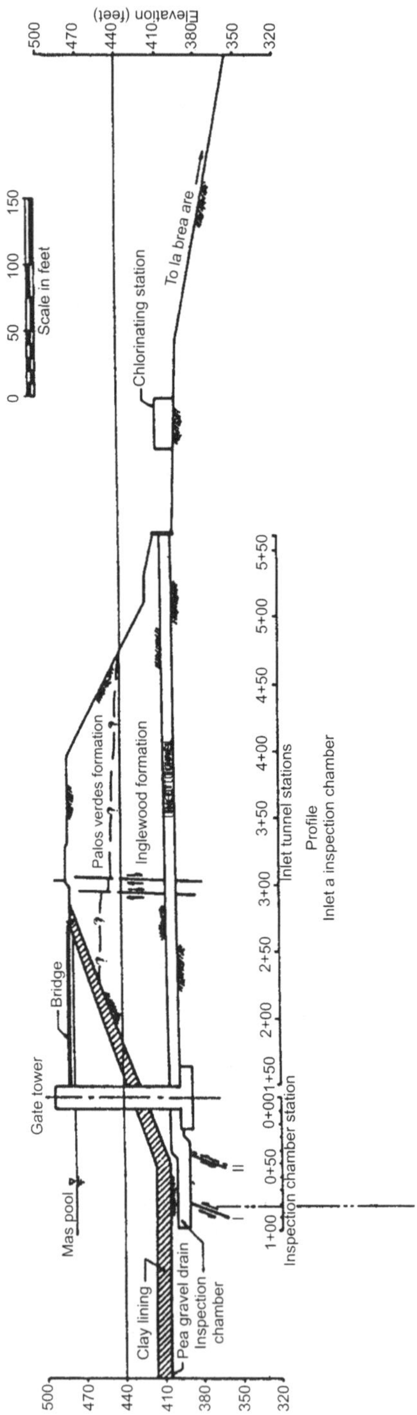

Fig. 2.8 Baldwin Hills Dam, California

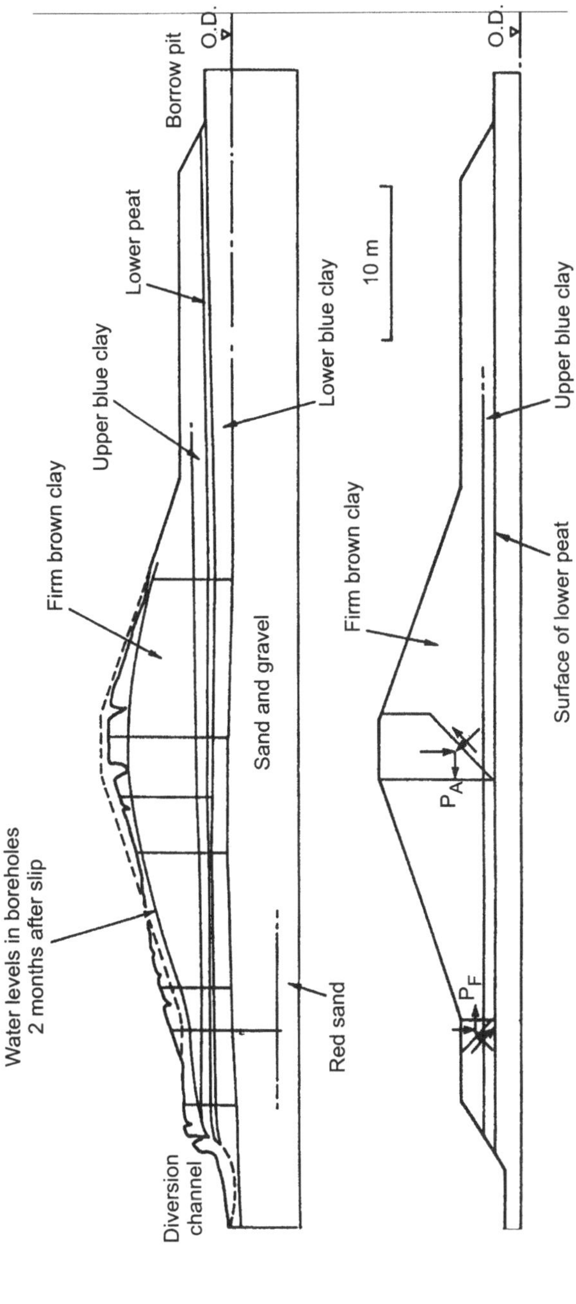

Fig. 2.9 Flood bank for the River Don at Thorpe Marsh

right side bank was to be made from material excavated from the channel. It was perfectly stable until the channel was excavated, thereby removing the lateral support from the soft blue clay, when it failed by the toe sliding towards the excavation. A total stress analysis gave F = 1.6 and it was suggested that the shearing resistance of the peat under and beyond the toe had been reduced by the redistribution of pore pressure along the layer from the higher construction values under the main body of the bank. This concept was checked by placing 12 hydraulic piezometers in the peat across the section of new bank to be built on the opposite of the channel. Measured values and details of the analysis were given by Ward et al. (1955). The failure was found to be along a soft layer in the foundation, but the failure surface was analysed by wedge shapes rather than by circular arcs.

Although a flood bank is a small structure in comparison with a large dam, its storage capacity can be the size of the ocean and failure could result in large areas of land being flooded. The importance of flood banks is often overlooked and, in general, they are not included in the reservoir regulations that apply to dams. This is a mistake the rectification of which is being considered in some countries.

Acu Dam

The A.R. Goncalves earth dam at Acu, R.G. Norte State in northeast Brazil is owned by the Brazilian National Department for Drought Regions (DNOCS), an organisation responsible for an area of about a million square kilometres. Since its formation in 1909, DNOCS has built about 265 large earth dams and more than 690 small ones, with an overall water storage capacity of more than 15 km^3. In some ways, DNOCS could be considered as the Brazilian equivalent of the United States Bureau of Reclamation (USBR) which also dates back to the beginning of the twentieth century and has built a large number of dams. After the initial design for the dam at Acu and the preparation of bid documents, there was a delay of about 3 years before bidding when tenderers were also invited for field inspection and design-as-you-go engineering. A further year elapsed before the beginning of construction. Independent geotechnic a consultants from Brazil and abroad were engaged to provide specialised services as might be required so as to modify the design according to the conditions encountered when the dam site was opened up and the conditions of the borrow materials could be more correctly assessed. Overall responsibility for the project was somewhat divided and there were some differences of opinion.

The dam was to be 40 m high with a fill volume of 5 km^3, to store 2.4 km^3, but during construction, when still 5.2 m below final crest level, a rotational slip type failure occurred on 15 December 1981, involving about 1.2 hm^3 of the fill. The slip surface began about 5 m downstream of the dam center-line in the surface of the placed fill, curving down to near ground level through the central part of the dam, then passing more or less horizontally to exit under the upstream toe. The construction surface fell about 15 m and the toe moved out 25 m. The failure from the first movement until it had come to rest, occurred in only half an hour.

The dam was founded on alluvial sand, a 22 m-thick overlying bedrock. The water table in the sand was close to ground surface, and it was intended to have a cut-off taken down to the bedrock. The initial design was planned to follow the practice at that time of the United States Bureau of Reclamation, as illustrated by Figure 15e of Wilson and Marsal (1979), shown here as Fig. 2.10. This type of cut-off was used for the Davis Dam in Arizona, Nevada, the Twin Buttes Dam in Texas, and many other USBR dams. At Acu, the ground water was going to be lowered by rows of well points on either side of the excavation, which would be cut with side slopes of 1 on 2 in order to provide stability in what should be dry sand, with if anything, a tendency for water to be drawn into the face of the slope, towards the pumped tube wells, thereby ensuring stability. As per estimate, it would require a year for the ground water lowering and excavation of the cut-off trench.

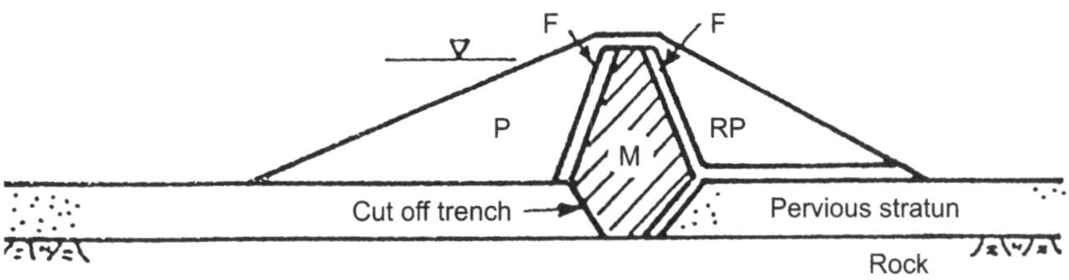

Fig. 2.10 U.S.B.R. type cut-off

There were essentially two main types of soil available locally that could be borrowed as fill. These were a dark grey-to-black flood plain silty clay, labelled BC, and an (unsaturated red terrace clayey sand and gravel, labelled RSC. Some parts of the borrow area showed a more gravelly red clayey sand, and this was labelled RSG. The below ground cut-off was to be formed by a wall of the low permeability silty clay, brought up from the exposed bedrock, supported by sand backfilled into the excavation of either side of the vertical cut-off as it was raised to ground level.

In order to save a year of construction, a counter proposal was made to move the below ground cut-off upstream to be just upstream of the dam toe, so indicated by Fig. 2.10, so that both cut-off construction and the placement of fill for the dam could go on at the same time. A 7m thick layer of the black silty clay (BC) was to be placed against the downstream I on 2 slope of the excavation through the sand layer, with the sand backfilled behind it. At the contact with the bedrock at the base of the excavation, the first layer of the clay was to be at a water content at about its liquid limit so that it would knit well into the rock surface to provide a water-tight seal. The next two layers were to be at a water content of Proctor optimum +2 to +4% and thereafter, the general water content specification was to be in the range of optimum ±1.5% to ensure a fairly soft and flexible clay that would not crack.

The black clay was also to be used for the core of the dam, supported by a downstream shoulder of the more gravelly red clayey sand (RSG), with a downstream zone of river gravel and rockfil. There was to be a vertical filter on the downstream face of the core, connected to horizontal filter drain placed at formation level. A section of this design is given by Fig. 2.11. It had been calculated that to protect the work during construction, a 14 m-high cofferdam would be required, and this was to be built as part of the upstream toe of the dam, using the red clayey sand as a fill that could be well compacted, and would be relatively impervious.

Having reached this stage, design discussion turned to the necessary instrumentation that should be provided to check on the safe behaviour of the dam—both during construction and subsequently while in operation with a full reservoir, and particularly during rapid drawdown. The advice of an international specialist consultant in North America was sought, and all details for the dam were sent to him. In addition to advice about instrumentation for measuring movements and pore pressures in the black clay core, the specialist suggested that the red clayey sand of the cofferdam would not be sufficiently impervious or of sufficient thickness near the toe in order to prevent reservoir water passing down into the foundation alluvial sands, and that the result would be difficulties. To avoid this situation, it was proposed that the 7 m thick layer of black clay forming the cut-off should be extended as a horizontal blanket over the alluvial sands to join the black clay core, as shown by Fig. 2.11(b), which could be better shaped within the dam body as a more traditional central core. To ensure flexibility with the idea of avoiding cracking, the placement water content of the black clay was to be in the range of optimum ±1.5% 0.

The dam was redesigned to produce the section shown by Fig. 2.11(b). The 14 m high cofferdam was to be built of the black clay and moved to be closer to the core so that it would form a part of the 7 m thick black clay impervious blanket connecting the core to the cut-off. By this time, the cut-off was well advanced and the rainy season was not far away. The contractor quickly built the long cofferdam, and to save time, used an

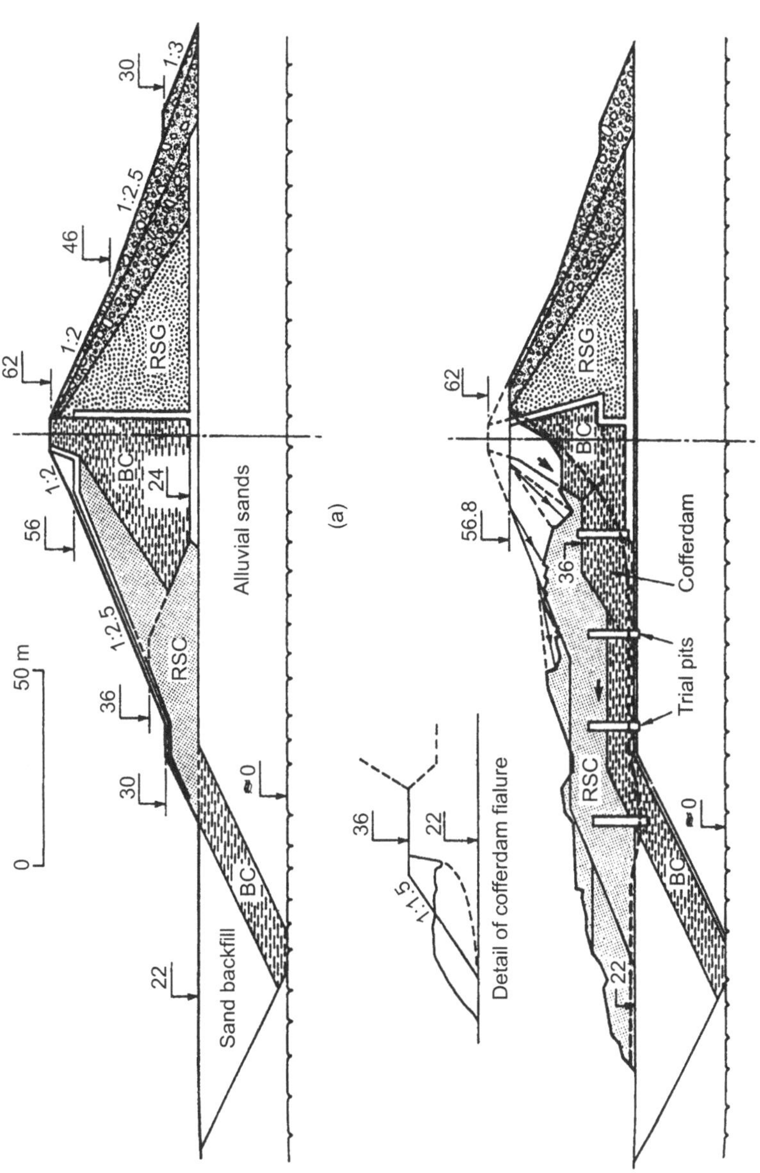

Fig. 2.11 Acu Dam: (a) original design and (b) the failed section

upstream slope of 1 on 1.5, as shown by the detail of the cofferdam in Fig. 2.11. It was to be supported at its toe by the 7 m thick impervious blanket so that a high factor of safety against a rotational slip was not required. After completion but before any of the toe weighting was added, two rotational slips occurred in the cofferdam, each 150 m long, involving about 160 × 103 m^3 of the black clay fill. These were in effect two full scale tests of the existing undrained shear strength c_u of the black clay and back analyses using total stresses, gave values of $c_u = 49$ kN/m^2. Unfortunately, this value was not used for a total stress analyses of the whole dam section. While waiting for design decisions, the contractor continued and rebuilt the long cofferdam to an upstream slope of 1 on 2.5.

While discussions continued about the final design of the dam, the contractor continued with construction to the section shown by Fig. 2.11(b). He placed the first layer of the black clay in a very wet condition to ensure flexibility, but later, the placement specification was changed to the range optimum –1.5% to +0.5%. As the geotechnical reader will be well aware, the use of percentages of water content above or below Proctor optimum can lead to very different conditions according to the types of soil being used as fill. A 2% change of water content has a much greater effect on a lean soil of low plastic index than it has on a fat soil. De Mello (1982) has pointed out that it is better to use a percentage of the Proctor value itself, say, 0.93 to 1.05 of the optimum water content. Because of the considerable time required to carry out Proctor test and also due to the variations that can occur in the values of Proctor optimum in different parts of a borrow pit, it has been found to be a better option to define placement water content on the undrained shear strength, c_u, of the fill. The value of c_u can be determined very rapidly on site with compression tests on tube samples taken from freshly placed and compacted fill. This approach has been described by Kennard et al. (1979).

A year after the two rotational slips in the black clay cofferdam, the contractor constructed the dam to within 5.2 m of the design crest elevation when a major rotational slip occurred. Eyewitnesses saw tension cracks developing in the downstream part of the core and shoulder fill, then, in the space of half an hour, the upstream part of the crest slid down 15 m while the berm and upstream toe moved out about 25 m. Trial pits and trenches showed that a rotational shear surface had been formed through the wide clay core and in the clay blanket, along the surface of the sand, in the layer of black clay that had been placed at the highest water content. Many faults were found in the design assumptions relating to the strength of the materials, including the effective stress parameters and the values allocated to pore pressures. Despite the long time span, the first piezometer had just been installed before the slip, but no readings had been obtained from it. Computer slip circles analyses had been made and on the upstream side, two end of construction circles passed only through the unsaturated red terrace clayey sand and gravel, giving factors of safety of 1.47 for a small 30 m radius circle through the toe, and 1.99 for a circle of 60 m radius passing through the shoulder fill but not cutting any of the soft black clay. A larger circle of 157 m radius passing from crest to toe—half of it through the black clay core and most of the rest in sand—was to represent the steady state condition giving a factor of safety of 3.6, or for rapid drawdown, 1.3. This factor had alluded the designers that the dam would be quite safe.

A back analyses of the dam failure, using total stresses, gave a value for the undrained shear strength of the black clay fill of $c_u = 48\ kN/m^2$, a value remarkably close that determined from the back analyses of the cofferdam failures. De Carvalho (1982) claimed that the back analyses revealed that the black clay had poorer geotechnical properties than had been expected, and that immediately after the 'accident', DNOCs invited the expert consultants Costa Nunes and Victor de Mello to report on the case. They suggested that the problem arose from an optimistic assessment of the geotechnical parameters for the black clay combined with high (unmeasured) pore pressures caused by high placement water contents. The Brazilian Government ordered restoration of the dam, based on a redesign entrusted to the Brazilian consultants. De Carvalho said, 'Professors Costa Nunes and Victor de Mello, with the help of a renowned staff, after studying the whole circumstances that caused the incident, prepared a new design whose features come to offer security and confidence'.

Carsington Dam

A reservoir was needed for the supply of an additional 225×10^6 l/d of water for a population of about 3×10^6 people in the East Midlands of Great Britain. A site was chosen in the valley of a small stream, which itself would provide some water, but primarily, the scheme was to be for pumped storage, extracting water from the River Derwent. The reservoir would cover about 295 ha at top water level. The dam was designed as an embankment dam that would be 1.225 km long to cross both the main valley of the small Scow brook and a subsidiary dip in the ground, going over a slight knoll between the two. When complete, the dam would carry a roadway that would act as a by-pass for the small downstream village of Hognaston.

To design and supervise the construction of the dam, the water company chose a well-established old firm of consultants who had been responsible for a large number of dams in the area. The old traditional design would have been for a puddled clay central core dam with an upstream slope of 1 on 3, and a downstream slope of 1 on 2 or 2½. The site was underlain by carboniferous mud stone that had weathered from a strong grey blocky material at depth to a soft yellow clay at the surface. The yellow clay, about 1 m thick, was not all in situ weathered; much of it had been formed by periglacial solifluction during the last glaciation. The downslope movement had developed shear surfaces, which could be found in trial pits by their gently undulating smooth surfaces. At one stage, the plan had been to use limestone from adjacent quarries for shoulder fill, but environmental objections were raised to the extension of the existing quarries and the heavy lorry traffic that would be produced. This caused the design to be directed towards using the less-weathered mudstone for the shoulders with the yellow clay and weathered mudstone for a rolled clay core.

Alterations occurred in the composition of the old consulting firm, with the introduction of new blood. Progress had brought the art and science of geotechnical engineering into dam design, and to bring the firm up-to-date, computers were introduced to help with stability analyses, and the modern idea of using drainage blankets within the fill to limit construction pore pressures, was brought into the design. Advice was also sought from independent geotechnical consultants who put forward the idea of leaving the yellow clay in place under the dam shoulders in order to increase the seepage path length through the mudstone under the dam. A grout curtain was to be formed slightly upstream of the core so that if further grouting were to become necessary, it could be carried out clear of the core itself and the roadway that was planned to run on the dam crest. To connect the core to the curtain, it was to have a base extension on the upstream side that would not only cover the grout cap but also unite with the existing layer of yellow clay. The resulting dam section is shown in Fig. 2.12. On the major section, where the dam crosses the Scow Brook, its height is 35 m.

A contract for dam construction was accepted in 1981 and preliminary works were carried out during the autumn, prior to the winter weather which tends to limit earthwork in that area. Dam construction began in April 1982 and by the end of the placing season towards the end of October, the fill level had reached about 18 m below final crest level. During the following placing season in 1983, the dam construction came to within 5 m of crest level.

There was considerable interest in construction, and various visits were arranged during the summer. The consultant prepared small brochures that were given to visitors describing the design of the dam and the work that could be seen. A group of civil engineering

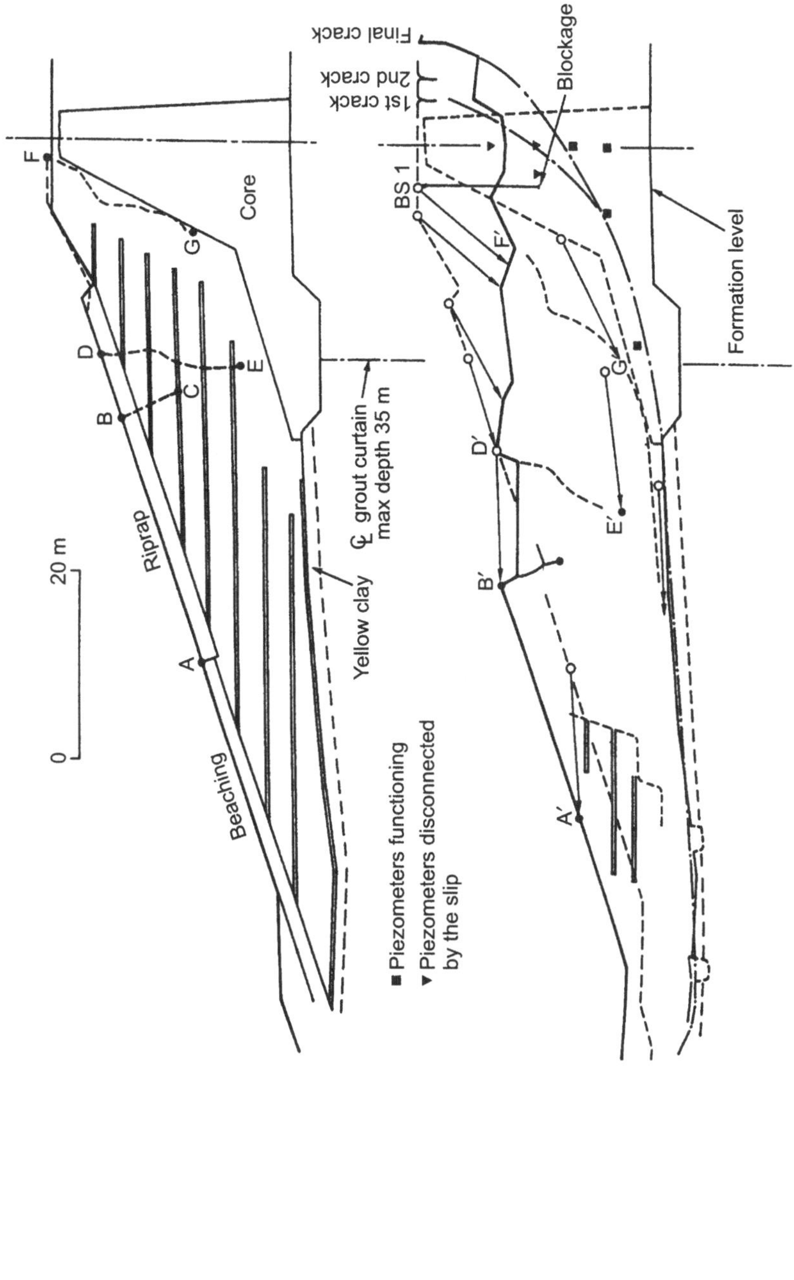

Fig. 2.12 Carsington dam: Upstream shoulder before and after failure

undergraduates, after such a visit, were asked by their supervisor to assess the stability of the dam from the details given by the brochure. They decided that the factor of safety was going to be remarkably low for the finished dam and that the probability of failure would depend on the rate of dissipation of construction pore pressures in the wide clay core and underlying layer of yellow clay. The contractor had his own geotechnical advisor, who also became concerned about the potentially low factor of safety, with the result that requests were made for a meeting with the consultant to review the design, but this was refused. The yellow clay had been eroded by the brook, so there was no layer of the soft yellow clay under the major section of the dam. As so often happens, stability calculations were concentrated on the major section and the circular slip surfaces of the computer programmes showed there to be adequate factors of safety for this highest section of the dam. Although the dam was well instrumented with piezometers in the core and shoulders, settlement gauges in the core and horizontal movement gauges in the downstream shoulder, where they would be able to be read after impounding, there were no measurement of horizontal movements of the upstream shoulder. Because of the concern about its stability, a berm was constructed at the upstream toe and reference points were installed in the upstream slope and datum points established on the valley floor upstream of the dam. Distances were measured by an electronic distance measuring instrument, with first readings taken on 18 August 1983. Fill placement continued until 3 October 1983, when work was stopped for the winter. Horizontal movements at the highest part of the dam amounted to 15.4 mm per metre height of additional fill. During the shut down period from 3 October 1983 to 10 April 1984, a creep movement of 35 mm was measured. The plot of horizontal movement against height of fill, given by Fig. 2.13, gives a clear indication that the horizontal movements were accelerating and, in fact, led to failure of the dam on 6 June 1984. Unfortunately, this plot was not drawn until after the failure, but it shows a method that could be used for future dams as an indication of potential stability.

Construction had continued under instructions from the consultant, although opposed by the contractor who was concerned about stability. On Friday, 1 June 1984, rain had prevented placement and on 4 Monday inspectors went on to the dam to see if placement could be continued. They found longitudinal cracks and work was begun to fill them in order to prevent the water getting into the fill, but movement continued and by Tuesday morning, the cracks were quite wide. Work began on raising the berm in a vain attempt to arrest movement, but during Tuesday night,

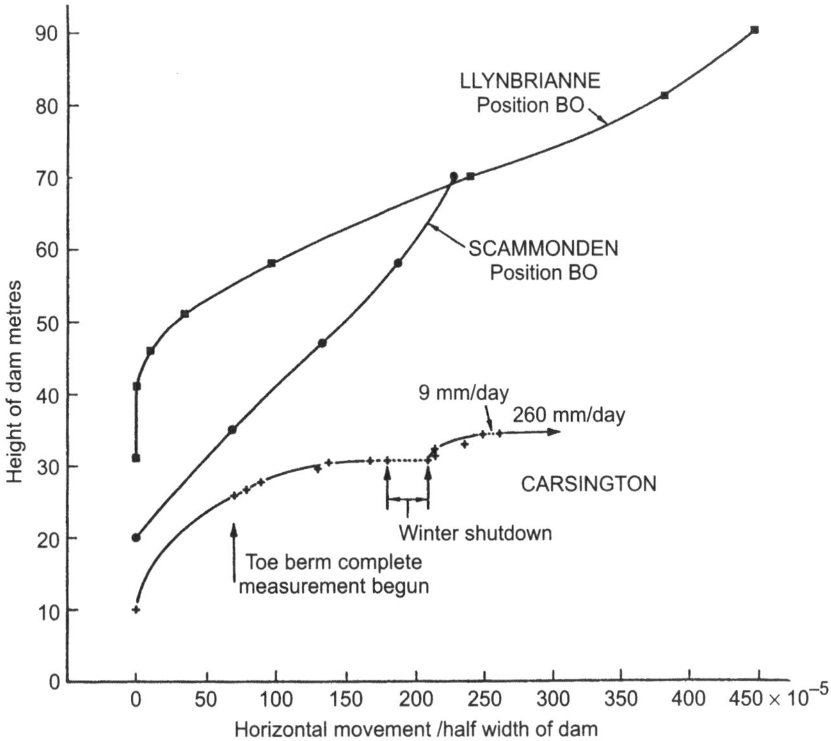

Fig. 2.13 Horizontal movement related to height of dam

larger movements developed and on Wednesday morning, 6 June 1984, the upstream part of the crest had sucked about 10 m and the upstream toe had moved out 13 m and risen about 2.5 m as it drove over the valley.

The failed dam is shown in Fig. 2.12 where it can be seen that the slip surface passed through the core and along the surface layer of soft yellow clay. Warnings had been given of the undesirable movements by an obstruction that had developed in the vertical settlement gauges; by the failure of a piezometer in the yellow clay that had leads crossing the slip surface and by measurements of settlement of the top of the core during the winter shut-down. These settlements are given in Fig. 2.14 and could be interpreted as showing the beginnings of the slip. In the analysis, some blame was put on the pre-formed slip surfaces on which strength could be expected to be at its residual value, but perhaps more important was the effect of progressive failure. The idea is that stain is far from uniform

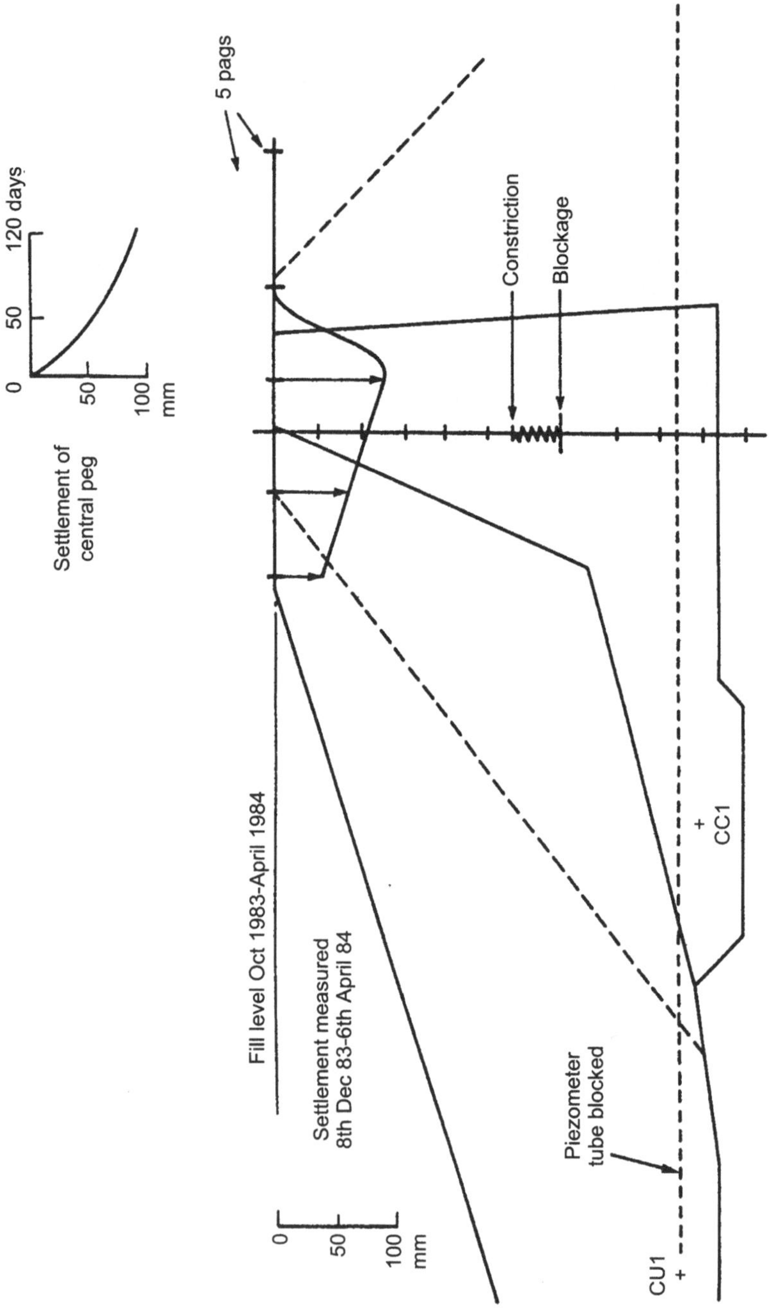

Fig. 2.14 Settlement of top of core and failed instruments

along the length of a potential slip surface and the clay could be taken over its peak to approach residual values in increasing stages as the final failure developed. This has been analysed by Potts et al., (1990) and this new method for analysing progressive failure may be considered as one of the more beneficial of the items to have resulted from the failure of this dam.

This dam failure that occurred during construction offered no risk to life or property because there was no water in the reservoir at the time. Geotechnical knowledge was adequate to provide a safe design and, in fact, showed the calculated factor of safety to be dangerously low. This dangerous situation was further revealed by the large measured settlements of the core during the winter shut-down and the failure of a piezometer and obstruction of a settlement gauge. Unfortunately, the responsible consultant dismissed these features, claiming that the use of different strength parameters would show an adequate factor of safety and that instrumentation in dams often fails so that there was no need to be concerned. Had the dam been designed in the traditional way with a central puddle clay core taken well down into the carboniferous mud stone and shoulders with slopes of 1 on 3 upstream and 1 on 2½ downstream, it could have been expected to be stable. As evidence of this, the downstream shoulder of the dam remained quite stable despite the failure of the upstream shoulder.

The failure has been very well described by various scholars (Coxon 1986; Skempton 1985; Skempton & Coats 1985; Skempton et al., 1989, 1993; Discussion 1994.

TAILINGS DAMS

Aznalcóllar Tailings Dam

It must be emphasised that the failure of Aznalcóllar tailings dam was not typical of tailings dams failures generally and it presented an interesting geotechnical problem. Unlike many tailings dams, this dam had been designed as a water-retaining dam, constructed from waste rockfill from the mine, with an upstream membrane of a clayey material, as shown in Fig. 2.15.

The site for the tailings impoundment that was to accept tailings from Los Frailes mine, 57 km west of Seville in Spain, was on the relatively flat floor of a valley on the right side of the Agrio river. The impoundment was rectangular in shape, about 1.2 km wide and 2 km long, extending from the river over the gently sloping ground back to the right side of the

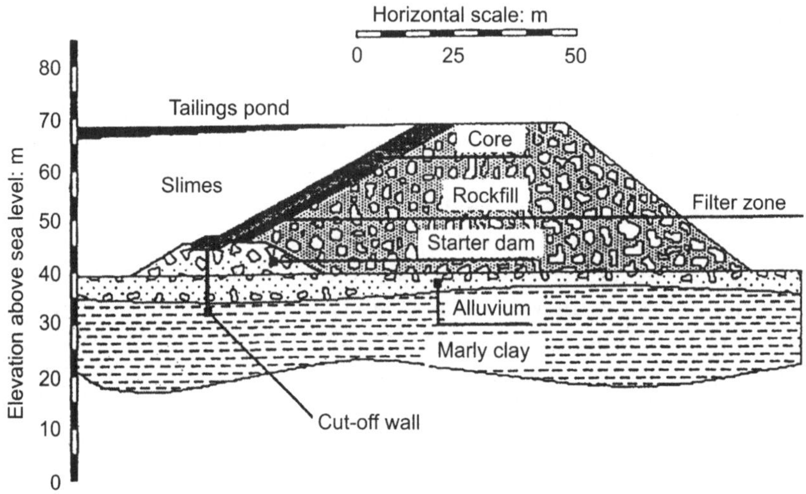

Fig. 2.15 Design section for Aznalcollar dam

valley where the ground began to rise more steeply. The 1.2 km long section of the dam extended from this valley side, across the plain to the river, then 2 km along the right side of the river, and back across the flatish land to the valley side again, with a dividing dam to separate two types of tailings. The river, on this fairly flat plain, meandered and over the long length of the dam adjacent to it, there were two meanders that came fairly close to the dam itself, the southernmost one progressively eroding ever closer to the dam toe.

The ground forming the foundation for the highest part of the dam, along the side of the river, consisted of about 4 m thickness of alluvium overlying a heavily overconsolidated clay, referred to as Guadalquivir blue marl of Andalucien age. This strong marl with values measured from borehole samples of 450 kPa, was clearly more than adequate to support the proposed 32 m high embankment dam that was to be built on the silty sandy gravelly alluvium and the designers had naturally no concern for the stability of the dam on such a good foundation. The stored tailings had quite a high bulk density of 28 kN/m^3, almost three times the weight of water. The dam, built from borrowed fill, was begun in 1978 and raised incrementally as the level impounded tailings increased. As so often happens during the long construction of tailings dams, the original ideal design section was not followed exactly: the section as built is shown by Fig. 2.16. After the second year, the upstream impervious membrane does not seem to have been constructed completely and as the dam approached full height, the river authorities began to complain about the

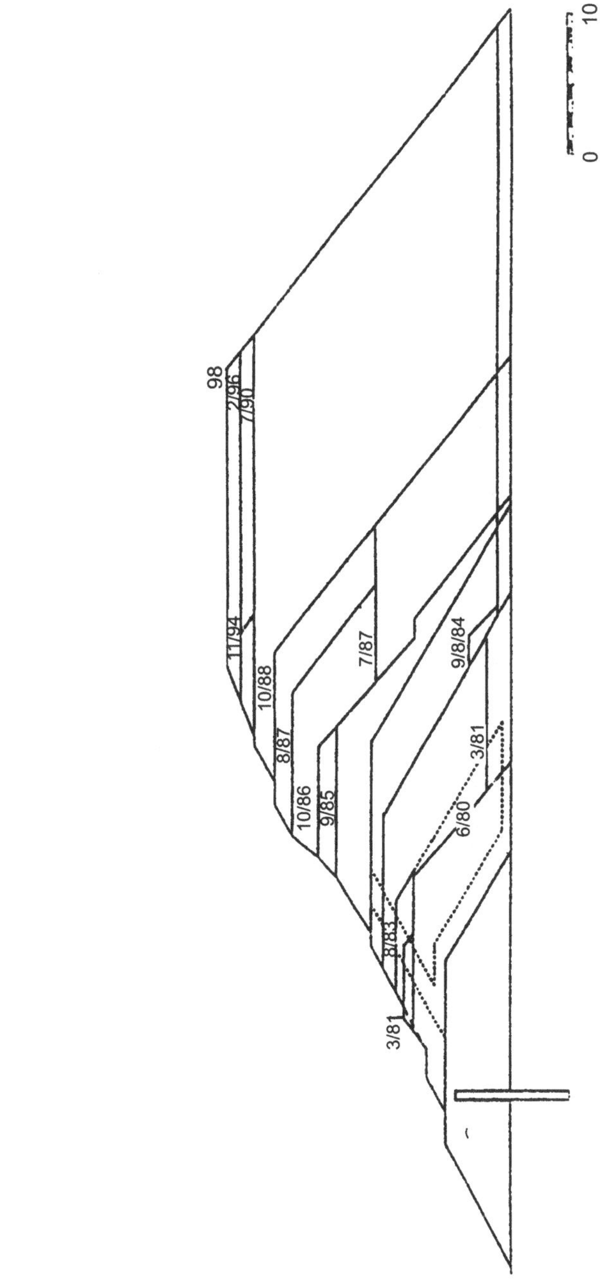

Fig. 2.16 As built section of Aznalcollar dam

amount of polluted water that was entering the river. A flow of 100 m^3/hour was measured and the mine owners attempted to reduce the leakage during 1995 by constructing a grout curtain from boreholes made into the dam crest over the length where the leakage was thought to be occurring. This step proved to be ineffective and, as a better solution, a line of 42 relief wells were excavated towards the end of 1997 between the dam toe and the river to catch the leaking water. They were 2 m diameter each, spaced at 25 m intervals and taken down through the alluvium to at least 1 m into the marl, backfilled with a selected sand and gravel. Several of these wells were inter-connected by thrust boring so that it was not necessary to have a submersible pump in every well. Water pumped from them amounted to 1000 m^3/hour. In 1996, the impoundment was subject to a full-scale stability study by independent experts and the Spanish authorities. No signs of instability were detected at that time; however it continued to be subject to regular third party inspections, the last less than two weeks before the failure without detecting the lack of stability that was rapidly approaching.

The dam had reached a height of 28 m when it failed during the night of 24 April 1998. The first indication was that the electric pumps in the relief wells failed and when the maintenance electrician went out to investigate, he heard the rush of water and found that the dam was releasing water and tailings from the impoundment into the river. A length of the dam of about 600 m had moved forward, and the end 60 m swung open like a door, the dam section remaining intact, releasing an estimated 7×10^6 m^3 of acidic tailings and water that flooded over an area of approximately 2.6×10^3 hectares of farmland. The breach occurred in the area where the meander of the river had eroded away the upper terrace to sculpt a steep scarp about 5 m high.

In order to study the cause of the failure, a site investigation was made, using 57 exploratory boreholes with a total length of 2117 m, and 27 piezometers plus 12 inclinometers were installed. It was concluded that constructional pore pressures had developed in the marl due to the ever increasing weight of the overlying tailings impoundment and the low permeability of the marl, combined with the large undrained area under the impoundment. It could be expected that the marl, contained some preformed slip surfaces, developed during earlier geological conditions, and it was presumed that the high pore pressures spread under the dam towards the river. As with the Thorpe Marsh failure, these high pore pressures developing under the unloaded ground beyond the downstream toe of the dam could considerably reduce the shear strength of the marl that was probably at its residual value, particularly on an ancient shear surface, which would also have helped to transmit the construction pore pressures laterally. The failure surface, as indicated by Fig. 2.17, was at some depth within the marl and passed under the river bed, raising the terrace several metres and discharging some blocks of marl, exposing a dark brown surface layer 2–3 cm thick on the marl where it had been in contact with the alluvium. This was thought to have been caused by the action of the acidic water from the impoundment that had a pH value of 3 to 4, attacking the carbonate content of the marl.

This failure is unique in the sense that the original designer in the 1970s could not reasonably have been expected to suspect this very hard, overconsolided marl that from the site investigation borehole samples appeared to be a homogeneous deposit with a more than adequate shear strength. More recent experience with progressive failure, knowledge of the large strength reduction of a brittle clay when subjected to even relatively small shearing strains and the presence of pre-formed shear surfaces and slickensides in some heavily overconsolidated clays might have indicated that a different design approach should have been used, but that is being wise after the event. Such considerations are given in the design of large water-retaining dams but it would be unusual for such details to have been applied to a small tailings dam.

Dam No. 26, Ohio River

The failure of the Aznalcóllar tailings dam was similar to that of Dam No. 26 in the sense that it was founded on a layer of hard shale. Although this small dam was only 3.4 m high, it failed on 8 August 1912 by sliding 47m downstream. A horizontal shear plane formed in the shale, it was thought, due to the pressure of water acting along an existing horizontal seam in the hard shale.

Stava

At 12.23 on 19 July 1985, two tailings dams—one above the other and both built by the upstream method—collapsed. A total quantity of 190,000 m^3 of tailings slurry was released and flowed, initially at a speed of 30 km/hr down into the narrow, steep-sided valley of the Rio Stava, demolishing much of the nearby small village of Stava and continued, at increasing speed—estimated to have been 60 km/hr—to another small town, Tesero about 4 km downstream, at its junction with the Avisio River. The only surviving eyewitness, a holiday maker, had the horrifying experience of watching the disaster from the hillside and saw the hotel where his family was having lunch being swept away by the

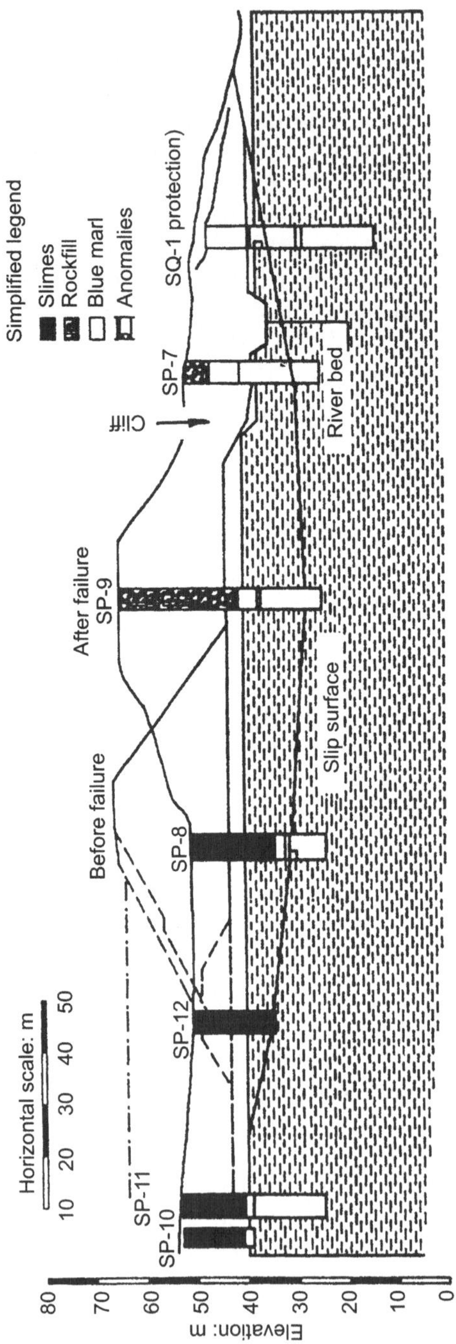

Fig. 2.17 Failure surface under the Aznalcollar dam

torrent of tailings. The damage caused by tailings is much more enormous than would be caused by the same flood of water, because the tailings are so heavy. Where water could flood a building, tailings can push it over and sweep it along with the flow. This failure caused 269 deaths.

The tailings dams, as indicated by Fig. 2.18, were for a fluoride mine begun in 1962 and were sited on a side slope of 1 on 8. The decant was in the form of a concrete culvert laid up the sloping floor, with coverable openings about every 0.5m vertical rise. Water from the pond decanted into the openings, which were covered—one by one—as the level of the tailings rose. The lower dam was built by the upstream method to a slope of 1 on 1.23. When it reached a height of 19 m, the second dam was begun at the upstream end of the impoundment and built to a slope of 1 on 1.43. When this second one reached a height of 19 m, further planning consent was required. This was given on the condition that a 5m-wide berm was constructed at that level and permission given for the dam to be built to a height of 35m. Construction continued at the same slope of 1 on 1.43 and the failure occurred when it was 29m high. The cause is thought to be due to a combination of blockage and leakage from the culvert under the toe of the upper dam, thereby raising the phreatic surface sufficiently to cause a rotational slip, as indicated by Fig. 2.18. Six months before the failure, a local slide occurred in the lower portion of the upper dam on its right side, in the area where the decant pipes pass underneath the dam, due to freezing the service pipe during a period of intense frost, according to Berti et al. (1988). For the next three months, water was observed to be seeping from the area of the slide. A month before the failure, the decant pipe underneath the lower impoundment fractured, allowing the free water and liquid mud from the pond to escape towards the Stava river, creating a crater above the point of fracture. A bypass pipe had to be installed through the top portion of the lower dam, and the broken decant pipe blocked in order to restore the use of the system. During this operation, the water level in the upper impoundment was lowered as far as possible. Then, just four days before the failure, both ponds were filled and put back into normal operation. 53 minutes before the failure, a power line crossing below the impoundments failed, then only 8 minutes before, a second power line also collapsed. The tailings from the failure reached Tesero about 4 km distance, within a period of 5 to 6 minutes. As a result of this failure, the strict Italian law governing the design and construction of water retaining dams, according to Capuzzo (1990), is being extended to include tailings dams.

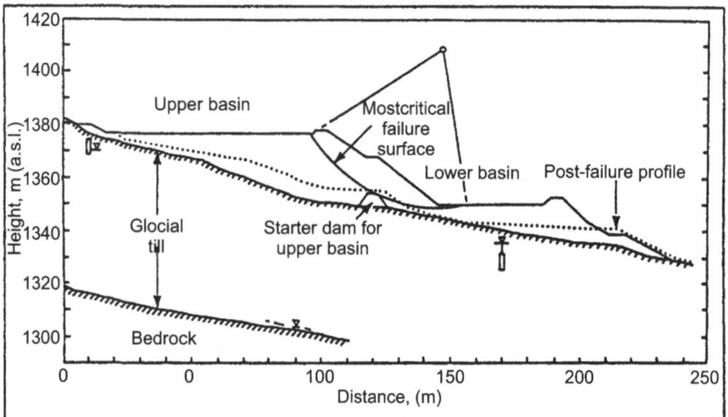

STAVA TAILINGS LAGOONS

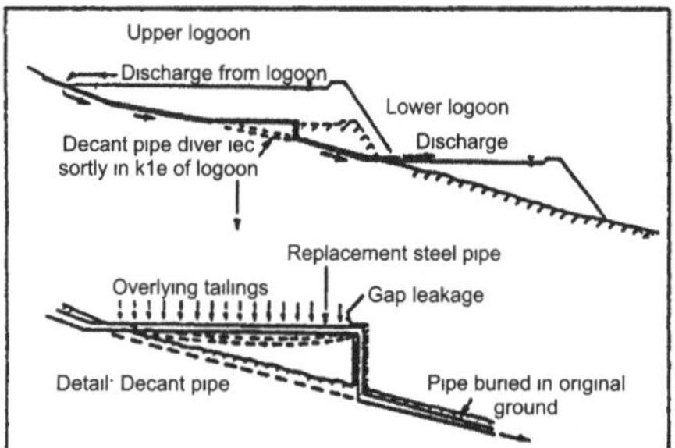

STAVA: FAILURE OF DECANT PIPE

Piezometric surface	Factor of safety
A (low)	1.45
B (intermediate)	1.30
C (intermediate)	1.12
D (high)	0.82

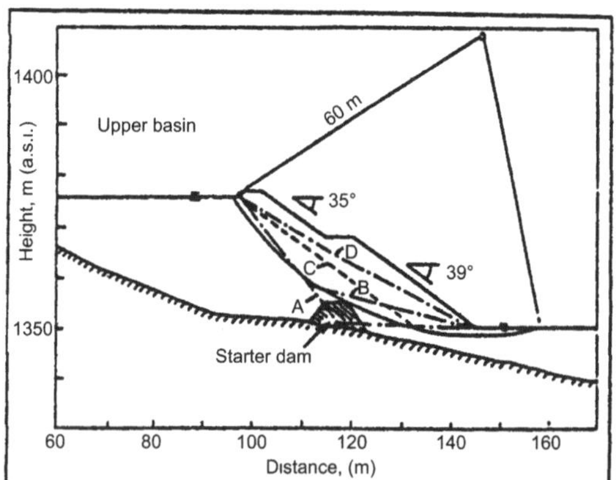

STAVA: PIEZOMETRIC SURFACES USED FOR STABILITY ANALYSES

Fig. 2.18 Stava tailings dams

Merriespruit

The Virginia No 15 tailings dam had been built by the 'paddock' method that is used extensively in South Africa's gold mining industry. It was a long dam encircling and retaining an impoundment of 154 ha. holding $260 \times 10^6 m^3$ of gold mine tailings containing cyanide and iron pyrite. The foundation soil was clay and drainage was required under the dams. General experience was that drains were often blocked by iron oxides and other residue. The impoundment formed one of several similar impoundments of the Harmony Gold Mine near Virginia in the Orange Free State. The suburb of Merriespruit containing about 250 houses had been built near the mine in 1956. Virginia No 15 lagoon was begun in 1974 and a straight northern section of the dam nearest the suburb was placed only 300m from the nearest houses. Dam construction and filling of the lagoon continued until March 1993, when the section of the dam closest to the houses was 31m high.

The summer of 1993/4 in the Orange Free State had been particularly wet and on the night of Tuesday, 22 February 1994, there were violent thunderstorms over Virginia and a cloudburst when 40mm of rain fell in a very short time. The water level in the lagoon rose due to direct catchment: there was no stream or other source of water that came into the lagoon, which while operational had a launder system that removed the transportation water decanted from the tailings slurry that had been delivered into the impoundment. During the early evening, at about 19.00, water was found running down the streets and through gardens and an eyewitness saw water going over the crest of the dam above the houses. The mining company and contractor were informed, but when their representatives reached the site, it was already dark. One of the contractor's men rushed to the decants and found water lapping the top rings but not flowing into the decants. He removed several rings to try to get the water flowing, but the main pool was next to the north dam crest with no direct connection to the decants. At the same time, another contractor's man was near the downstream toe of the dam, and saw blocks of tailings toppling from a recently constructed buttress that had been built against a weak part of the dam. An attempt was made to raise the alarm, but before anyone had been contacted, there was a loud bang, followed by a wave of liquefied tailings that rushed from the impoundment into the town. Cross-sections of the dam during early stages of failure, given by Blight (1997) are shown in Fig. 2.19.

A 50m-wide breach formed through the dam, releasing $2.5 \times 10^6 m^3$ tons of tailings that flowed for a distance of 1,960m, covering an area of

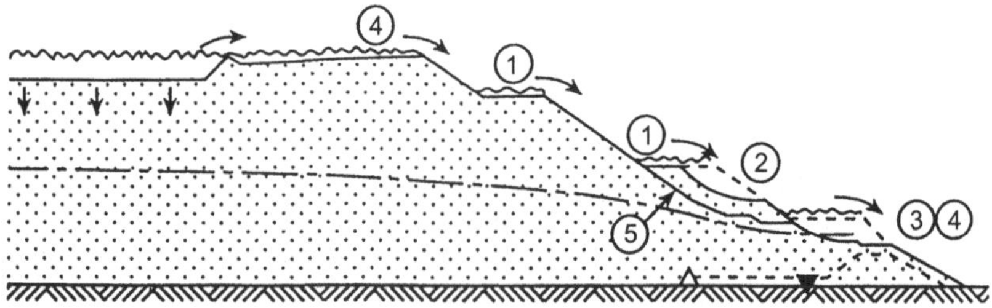

CRITICAL SECTION OF NORTH WALL DURING EARLY STAGES OF FAILURE

1. Berms overtop after thunderstorm.
2. Loose tailings infill to earlier failures on lower slope erodes.
3. Tailing buttress starts to fail.
4. Pool commences overtopping and erodes slopes and tailings buttress.
5. Unstable lower slope fails and failed material is washed away.

Fig. 2.19 Section showing probable failure mechanism

$520 \times 10^3 m^3$. The flow passed through the suburb where the power of the very heavy liquefied tailings demolished everything in its path—houses, walls, street furniture and cars, carrying people and furniture with it. According to newspaper reports, people already in bed at about 21.00 hours when the mudflow struck, found themselves floating in their beds against the ceiling. 400 survivors spent the night in the Virginia Community Hall, a kilometer away. Hetta Williamson said that her husband had gone back in daylight to their former home and found nothing but the foundations. It is remarkable that only 17 people were killed in this disaster.

Apparently, this north section of the enclosing dam had been exhibiting signs of distress for several years, with water seeping and causing sloughing near the toe. A drained buttress constructed from compacted tailings had been built against a 90 m-long section, but continued sloughing had caused the mine to stop, putting the normal flow of tailings into the impoundment more than a year before the failure, i.e. the impoundment had been closed. At that time, the freeboard was—according to the contractor—a respectable 1 m. But sloughing at the toe continued, and construction of the buttress was continued. Not long before the failure, slips had occurred in the lower downstream slope just above the buttress. In fact, although the placement of tailings had been stopped, wastewater containing some tailings continued to be placed and the water overflowed into the decants. Unfortunately, there formed a sufficient deposit of further tailings to cut off the decants and cause the main pool of water to move towards the crest of the north part of the dam, leaving a freeboard of only 0.3 m, and water was still being pumped into the impoundment from the mill on the night of the failure. Evidence of what had been going on since supposed closure was provided by satellite photography. A Landsat satellite passed over the area every 16 days and the infrared images revealed the positions of the tailings and the water pool.

Government Mining Regulations that had come into force in 1976, required a minimum freeboard of 0.5m to be maintained at all times for this type of impoundment, to enable a 1 in 100 year rainstorm to be safely accommodated without causing overtopping. Evidence of the level of tailings in the Virginia No 15 lagoon showed that the tailings had been brought up to within 15cm of crest level prior to abandoning this storage in March 1993. Had the government regulations required inspection of the dam, particularly at closure, the very small freeboard would have been noticed and a further raising of the dam crest enforced to prevent overtopping in the event of a maximum probable precipitation.

DAMS DAMAGED DELIBERATELY DURING WARS

The 63 m high Peruca rockfill embankment dam has a relatively narrow central clay core. It was completed in 1960 and it had been regularly inspected until 1991 when the Patriotic War in Croatia caused disruption. The Serbian army gained access to the dam on 28 January 1993 and placed explosive charges in the grouting gallery and in the spillway. As a result of international concern, the dam had been taken over by United Nations' Military Observers in October 1992. Captain Gray of the British Army who was one of the Observers, on realising the potential danger, raised the spillway flaps, thereby reducing the reservoir level by 5 m. The explosive charges were detonated at 10.48:55 on January 28, 1993, an event recorded by three seismic stations some distance from the dam. From the seismic records, it was calculated that the total weight of explosives was the equivalent of 20 to 30 tons of TNT. A crater of 25 to 30 m diameter and 10 m deep was formed in the crest. Water flooded over the upstream rim, but fortunately the downstream rim remained 30 to 50 cm above the water level. Had that level not been lowered by 5 m, overflowing and a breach would have occurred. The Croatians were fortunate not only that the reservoir had been lowered, but that the saboteurs did not seal the entrances to the gallery. The explosive force dissipated to the extent that the vegetation was damaged in front of the openings over a distance of 150 m. The reservoir was subsequently emptied in a controlled fashion spread over some time, and the dam has been repaired.

In Russia, the 43 m high curved concrete gravity Dnjeprostroj dam was constructed during the period 1928 to 1931 across the great Dnjepr River near the city of Saporoshje to serve a hydro-power plant that was the largest in the world. It had a crest length of 762 m, carrying 47 spillway gates. In September 1941, Soviet troops blasted a part of the dam with dynamite during their retreat from the German army. A breach 200 m wide was formed and the water escaped at a rate of 35000 m^3/sec. Immediately after their arrival, the Germans began reconstruction and by the end of 1942, the breach was closed. Nine months later, the tide of war having reversed, the Germans attacked the dam from the air, and the bombs took out the repaired section. After the war, the Russians reconstructed the dam, which today is fully operational.

The Mohne masonry gravity dam in Germany on the Möhne River, a tributary of the Ruhr, was built in 1908–1913 to a height of 40 m in order to supply water of the densely populated Ruhr valley. It was attacked by the British air force, using the specially developed bouncing bomb on 17 May 1943, and a 77 m wide and 23 m deep breach was formed through which the water poured at an initial rate of about 8000 m^3/sec. In the small Möhne valley downstream, the flood wave reached a height of 10 m and caused the deaths of 1200 persons. The upstream face of the dam had been protected against bomb or torpedo attack by nets hung down from floats. The design of the bouncing bomb was that it would skip over the surface of the reservoir, thereby jumping over the nets until it hit the dam, when it sank to be detonated at a specific depth by water pressure. The raid was not without considerable loss to the British air force.

At the same time, the 47 m-high Eder masonry gravity dam was attacked in the same way. A breach 50 m wide and 25 m deep was blown in the dam, releasing a flood that caused serious damage in the Eder and Fulda valleys and the loss of 400 lives. An attack was also made on the 60 m-high Sorpe embankment dam, but this proved to be more resistant and apart from a 12 m deep bomb crater, it remained intact.

FAILURES ON LINE OF PIPES AND CULVERTS

The 29 m-high Dale Dyke embankment dam described above was built with twin outlet pipes passing through it. They were spigot and socket cast iron pipes, 45 cm diameter in 2.75 m lengths. The sockets were 15 cm long and the spigot of the next length was sealed in by caulked lead. They were laid in a trench about 2.7 m wide and deep below formation level under the dam. After the failure, Government Inspector Rawlinson said that no waterworks embankment should have a line of pipes laid through it in such a way as to preclude any possibility of inspection or repair. He considered that failure was probably due to leakage from a fractured pipe causing subsequent erosion. As has been described above, we now know that this was not the cause of failure, but the result of his opinion given in 1864, was that no further use was made of pipes passing through an future British designed embankment dam.

Although pipes as such are not put through embankment dams, there is a tendency to place culverts through them. The better practice is to tunnel through the bedrock of an abutment for the outlet, culverts through the dam itself have been used. But they do fail and we will look into two recent examples.

A second dam on the Wadi Gattara in Libya built 1969 to 1970 was a part of a flood protection scheme. It was put into operation in 1972, but the reservoir never filled, which was perhaps as well since the purpose of the dam was to mitigate flooding. But on 21 December 1977, the dam failed when the water level was only about third full height, by wash out along the line of the culvert. The dam, about 14 m high, was of homogeneous construction with a bottom outlet culvert designed to take an outflow of 35 m^3/sec. passing through the fill.

A further example was that of the Zoeknog dam on the Mutlumuvi River near Bosbokrand in the Northern Transvaal Region of the Republic of South Africa. It was a 38 m high dam of homogeneous construction with a centrally positioned morning glory spillway, connected to a 102.7 m-long horseshoe-shaped concrete culvert 9 m high and 9 m wide, passing through the dam body and terminating in a stilling basin at the downstream toe. The upstream slope of the dam was 1 on 2.5, flattening to 1 on 4 below the lowest berm. The downstream slope was 1 on 2 to 16 m below crest level, then 1 on 2.5. Impounding began in November 1992 and by the third week of January 1993, the reservoir level was 18.6 m above the river bed, i.e. 64% of full reservoir height. The failure occurred when it was in this condition. To begin with, water was seen flowing freely from the embankment a long the left side of the conduit. A few hours later, a large hole was seen to have formed from the dam's foundation level to almost half the height of the dam. Shortly afterwards, the crest collapsed into the breach and about an hour later, the breach washed right through, releasing the reservoir. A section of the spillway and culvert are shown in Fig. 2.20. It was concluded that failure had been initiated by hydraulic fracturing associated with water pressures that exceeded soil pressures in or immediately above the excavation trench adjacent to the conduit. These hydraulic fractures caused excessive seepage, leading to piping attributable to erosion loss of embankments material.

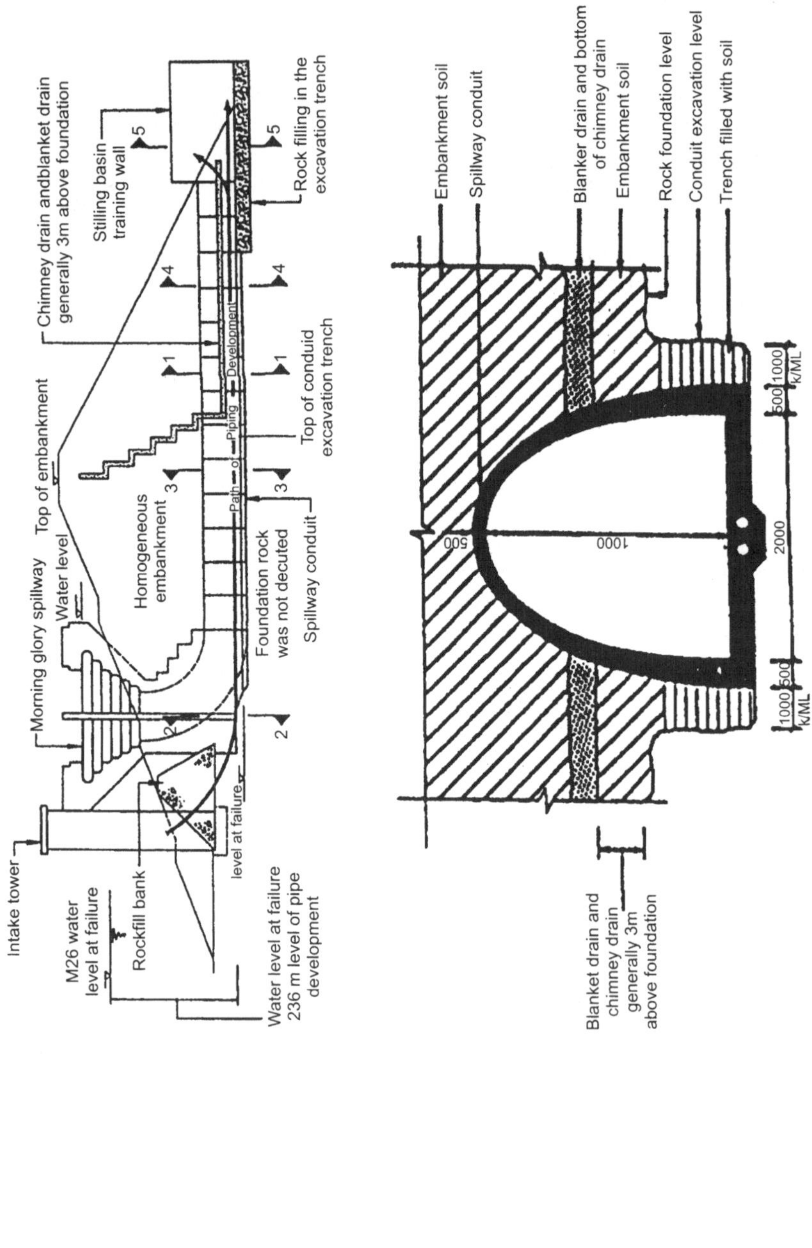

Fig. 2.20 Section of dam and culvert: Zoeknog dam

REFERENCES

Aswathanarayana, G., and Subrahmanyam (1998). Rehabilitation of Dams - Case Study in Karnataks' *Proc. Of Symposium of Rehabilitation of Dams,* New Delhi, pp. 192–200.

Benson RP, Harland WP and Pinkerton L (1983). The ancient Maduru Oye sluiceway. *Int. Water Power and Dam Construction,* Vol. 35, no 12 (December), pp. 26–31.

Berti G, Villa F, Dovera D, Genevois R and Brauns J (1988). The disaster of Stave/ Northern Italy. *Proc. Specialty Conf. "Hydraulic fill structures"* GT Div/ASCE, Fort Collins, Co, pp 492–510.

Berga L (1997). Failures and hydrological incidents of dams in Spain. *Trans 19th Int Congress Large Dams,* Florence, Vol. 4, pp 417–430.

Blight GE (1997). Insight into tailings dam failures—Technical and management factors. *Proc. Intn. Workshop on Managing the Risk of Tailings Disposal,* Stockholm, Sweden, May, pp 17–34.

Capuzzo D (1990). A brief description of the main causes underlying the collapse of Stave tailings dams, Trento - Italy. Proc. ICOLD Intn. Symp. Safety and Rehabilitation of Tailings Dams, Sydney, Australia, vol 2, pp 64–66.

Chandler RJ (1991). Slippery slopes and swelling soils. *Inaugural Professorial Lecture, Imperial College of Science,* Technology and Medicine, London.

Coxon RE (1986). Failure of Carsington embankment. *Report to the Secretary of State for the Environment,* HMSO, London. 236 pp.

De Carvalho LH (1982). Contribution to discussion on Question 55. *Trans 14th Int Congress on Large Dams,* Rio de Janeiro, Vol. 5, pp 551–554.

De Mello, Victor (1982). A case history of a major construction period dam failure. *Amici et Alumni.* Em. Prof. Dr Ir E.E. De Beer. pp 63–78.

Discussion on Carsington (1995). *Géotechnique* Vol. 45, no 4, pp 719–739.

Jansen RB (1980). *Dams and Public Safety.* Publ. by U.S. Dept of the Interior. 331 pps.

Kennard MF, Lovenbury HT, Chartres FRD and Hoskins CG (1979). *Shear strength specification for clay fills.* Clay Fills. Instn Civil Engrs, London. pp 143–147.

Leonards GA (1987). Dam Failures: *Proceedings of International Workshop.* Pub. Elsevier, Amsterdam, 612 pp.

Mande, O.D., Sinha, D.K., and Harijan, S.D. (2000). Causes of Distress/Failure of Dams, *Proceedings of 3rd International R&D Conference,* Jabalpur, India, pp. 236 and 237.

Mistry, J.F. (1996). Modes and Various Causes of Dam Failures with Case Studies; *Proceedings of Workshop on Dam Safety,* Jaipur, India, pp. V-1 to V-17.

Morris SC (1998). *The Crucible of Creation.* Oxford University Press, 242 pps.

Murti, N. G. K. (1978). 'Earth Backing of Masonry Dams', *Irrigation and Power Journal,* Central Board of Irrigation and Power, New Delhi, April, pp. 189–199.

Murti, N .G. K. (1979). 'State of Art of Earth Backing of Masonry Dams', *Irrigation and Power journal,* Central Board of Irrigation and Power, New Delhi, October, pp. 351–366.

Murthy, Y.K., Mane P.M. and Pant, B. (1979). 'Tensile Failures in some Stone Masonry Gravity Dams in India', *Proc. ICOLD,* New Delhi, Q. 49 R.31.

Olalla C and cuellar V (2001). Failure mechanism of the Aznacollar Dam, Seville, Spain. *Geótechnique* Vol. 51, no 5, pp 399–406.

Pearson R, Langrick H and Underhill A (1982). Sources on archaeological hydraulic systems and sluiceways. Report prepared for Crippen International Ltd., Concept Canada Ltd. and Klohn Leonoff Ltd, on behalf of CIDA.

Penman ADM and Milligan V (1993). Longevity of embankment dams - A critical review. *International Workshop on Dam Safety Evaluation,* Grindelwald, Switzerland, Vol. 1, pp 33–51.

Pessoa JC (1982). New construction methods and materials. *Trans 14th Int. Congress on Large Dams,* Rio de Janeiro, Vol. 5, pp 680–682.

Potts DM, Dounias GT and Vaughan PR (1990). Finite element analysis of progressive failure of Carsington embankment. *Géotechnique,* Vol. 40, no 1, pp 79–101.

Quintela AC, Cardoso JL and Mascarenhas JM (1987). Roman dams in southern Portugal. *Water Power & Dam Construction,* May issue, pp 38–49 & 70.

Rao KL (1951). Earth dams, ancient and modern in Madras State. *Trans 4th Int. Congress Large Dams,* New Delhi, Vol. 1, pp 285–301.

Simek M (1997). Bilá Desńja dam failure. 80th Anniversary. *Trans 19th Intn. Congress on Large Dams,* Florence, Vol. 4, pp 85–91.

Skempton AW (1985) Geotechnical aspects of the Carsington Dam failure. *Proc. 11th Int Conf Soil Mech,* San Francisco, Vol. 5, pp 2581–2591.

Naqvi, S.R.A. And Parashar, Ashop (1998). 'Failure and Remedies of Chandora Dam in Madhya Pradesh' *Proc. of Symposium on Rehabilitation of Dams,* pp. 466–46.

Prakash Chand, Jaya, (1998). 'Under Water Treatment of Cracks in Hirakud Dam', *Proc. of Symposium on Rehabilitation of Dams',* New Delhi, pp. 244–255.

Sakhalkar, S.C. and Shahane V.C. (1989). 'Failures And Distress in some of the Dams in Recent Years in Maharashtra', *Proceedings of 55th R&D session,* Technical Session VI, Central Board of Irrigation and Power, Aurangabad, India.

Sundariya, E. and Kalsi, R.K. (1996). 'Dam Failures - Some Case Studies', *Proceedings of Workshop on Dam Safety,* Jaipur, India, pp. V-25 and V-26.

Skempton AW and Coats DJ (1985). Carsington Dam failure. Failures in earthworks: *Proceedings of Symposium.* Thomas Telford, London, pp 203–220.

Skempton AW, Norbury D, Petley DJ and Spink TW (1989). Solifluction shears at Carsington, Derbyshire. *Proc. 25th An Conf Engng Gp,* Geol Soc, London, pp 277–285.

Skempton AW and Vaughan PR (1989). Discussion on Carsington Dam. In: *Clay Barriers for Embankment Dams,* pp 182–186. Thomas Telford, London.

Smith NAF (1970). *The Heritage of Spanish Dams.* Spanish National Committee on Large Dams, Madrid. 176 pps.

Smith NAF (1971). *A History of Dams.* Peter Davies, London. 279 pp.

Smith NAF (1975). Man and Water. *A history of hydro-technology.* Peter Davies, London. 239 pp.

Van Loon, H (1922). *The Story of Mankind.* George G. Harper & Co Ltd, London. 480 pps.

Vogel A (1993) Research for risk evaluation of dam failures by defining a break-dimension-time-factor. *International Workshop on Dam Safety Evaluation,* Grindelwald, Switzerland, Vol. 1, pp 21–31.

Ward WH, Penman ADM and Gibson RE (1955). Stability of a bank on a thin peat layer. *Géotechnique,* Vol. 5 no 2, pp 154–163.

Wilson S.D. and Marsal R.J. (1979). Current Trends in Design and Construction of Embankment Dams. American Society of Civil Engineers, New York. 125 pp.

Chapter 3

Misconceptions in the Design of Dams

INTRODUCTION

The concept that all dams must contain a waterproof element supported by a stable structure is often misunderstood. Design methods have continued to advance as a part of engineering progress. Conceptions accepted in one age may be considered unacceptable in another. Advocates of the most modern ideas feel closer to an absolute truth than their predecessors. At any stage, earlier concepts may be scorned as misconceptions. At any one time, however, there are invariably a variety of conceptions in current use and some will be considered to be misconceptions by those who hold other views.

GRAVITY DAMS

The design principles published only seventy seven years ago by Wegmann (1927), presumably representing the conceptions accepted at the time, would now be dismissed by all designers as misconceptions. In discussing drainage in gravity dams, Wegmann wrote:

In Vyrnwy dam in North Wales, a system of drains was constructed in the foundation to relieve the base of the dam from upward pressure that would occur if water leaked under the dam.

'It is doubtful whether a system of drainage placed in the dam or its foundations is an advantage, except in special cases. By giving free outlet to the seepage water, such a system encourages leakage. Most engineers prefer to construct dams as tight as possible so as to offer the greatest resistance to seepage. Water that passes through a thick masonry dam and only shows itself as a moist spot on the downstream side can scarcely be considered to cause an upward pressure in the masonry. On the contrary, in passing through the pores of the stones and mortar, it increases the weight of the masonry and, consequently, the stability of the dam'.

This quotation shows lack of appreciation of the simple concept that all dams must contain two basic features: a waterproof element and structure that will support it safely.

In the past, homogeneous sections of both earthfill and masonry have been expected to perform both functions. Prior to the catastrophic failure of Bouzey dam (20m) on 27 April 1895 (ICOLD 1974), gravity sections had been designed by taking account only of reservoir water pressure on the upstream face and the weight of the dam, limiting compressive stress to about 600kN/m^2. Improved cements enabled allowable stress to be increased to 1000kN/m^2 but the resulting reduced thickness of the section was enough to produce instability under the previously discounted uplift pressure. Maurice Lévy (1895) proposed that a design should ensure that the vertical compressive stress in the masonry at the upstream face should exceed the pressure from the reservoir water at all levels in order to prevent ingress of water into the body of the masonry. While this approach would avoid the catastrophe of overturning under uplift pressure, it would not prevent percolation of water through the dam, with accompanying risk of damage by acid water, staining and frost damage at the downstream face. A solution, which has not yet been accepted by all designers, is to provide an 'impervious' membrane on the upstream face, with drainage behind it. A way of achieving this concept was to construct the upstream face of the dam as a continuous series of jack-arches. Leakage through the flat outer surface above the arches would drain into the vertical spaces under the arches, down to the base of the dam where water could be carried through into drainage galleries in the body of the dam.

The bodies of masonry dams can be relatively pervious and usually the upstream face is intended to be fairly impervious and acts as the waterproof element. Stone masonry dams in China (CHINCOLD 1979) have been built mainly since 1949, although an early example, built in 833 on the Daixi river in Zhejiang Province, is 27m high and still in service. The post - 1949 stone masonry dams generally use rubble masonry laid in a low grade cement mortar. In practice, it is usually impossible to completely fill up all the gaps and joints, so that the mass forming the body of the dams tends to be fairly pervious and to control leakage, an impervious facing is made from ashlar masonry (or processed rubble masonry) laid in high-grade cement mortar. Alternatively, a concrete facing may be placed on the upstream face of the dam. Both these methods give the desired effect of an impervious upstream slab, supported by a fairly free draining body strong enough to support the thrust from the pressure of the reservoir water.

As a remedial measure, percolation through the 33m-high Ringedals gravity dam in Norway was stopped when it was about 10 years old

(Groner 1933) by building a completely separate reinforced concrete slab as a waterproof element, supported from the original structure of the dam. The slab was made 0.47m thick at the base, reducing to 0.2m at the top and was held 2m away from the upstream face of the original dam (Fig 3.1) by struts at 2.25m spacing horizontally, with vertical spacing varying from 1.8m at the bottom, up to 3m near the top. The 2m gap was drained from its base.

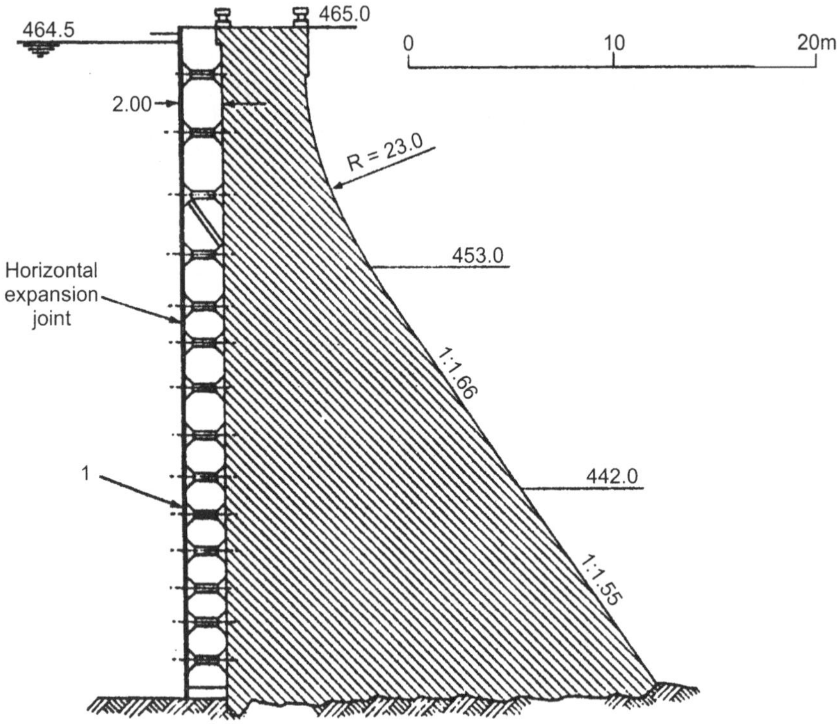

Fig. 3.1 Ringedals gravity dam

The Hong Kong 48m high Shing Mun gravity dam (Fig 3.2) followed this concept and was designed to have separate slab supported from the main body of the dam by precast blocks (Binnie 1936). Several Italian dams had faces composed of a succession of jack-arches with vertical support walls at about 2m centre to hold them off the main body of the dam and allow draingage. The 52m-high gravity Ceresole dam built towards the end of the 1920s, was provided with a separate concrete face made up of jack-arches to carry away any leaking water. This has been described by Scuero and Vaschetti (1997) as a Levy type facing that would

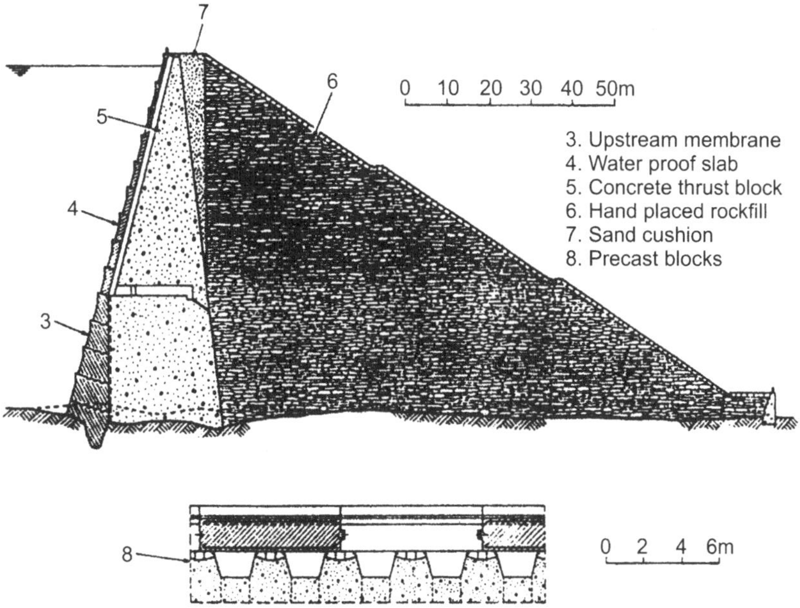

Fig. 3.2 Shing mun gravity dam.

keep the main body of the dam relatively dry. Twenty years after construction, the facing was leaking sufficiently to cause some remedial measures to be carried out in the form of filling the jack-arches with concrete. A few years later, a new face was built of stone masonry, with cement mortar joints to form a continuous layer of low permeability. About another 15 years later, leakage had increased again to such an extent that dangerous uplift pressures developed in the lower right side of the dam, and further remedial measures became necessary. After considerable consideration, it was decided to excavate a new drainage gallery low down in the upstream part of the dam, install a new grout curtain and then waterproof the upstream face with some sort of impervious covering. Coatings of mortar and resin were considered, but since the use of PVC sheets had become established for other dams, and knowing the weaknesses of the older types of covering which were likely to be pushed off by water pressures from within the dam body in the absence of drainage immediately under the covering, PVC became an attractive option. The application of this remedial measure is described below.

This use of separate waterproof faces can be regarded to be an ideal solution. In mass concrete gravity dams, a more generally used solution is to provide drainage through horizontal galleries interconnected by

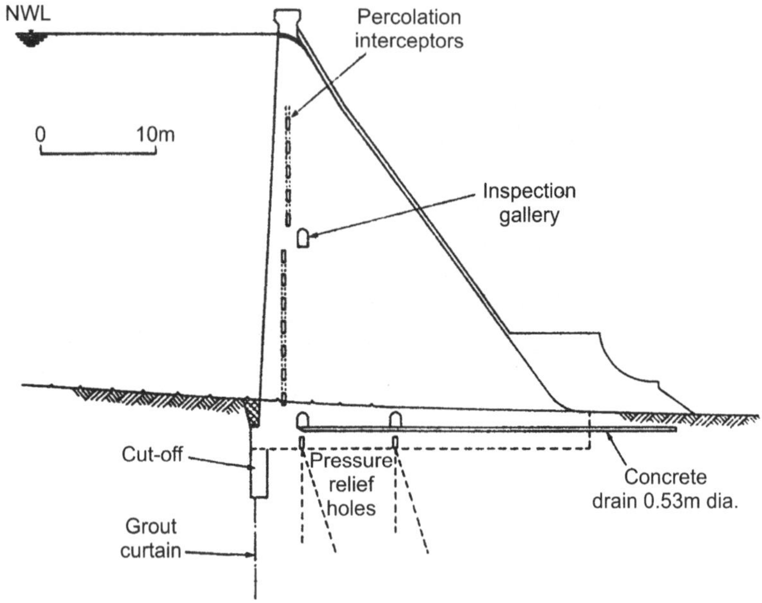

Fig. 3.3 Boothwood dam

vertical holes at close centers, all lying on a plane a short distance from the upstream face. A richer concrete is used for the 'slab' upstream of the drains, than that for the remainder of the dam. An example of this approach is shown by Boothwood dam (59m) in England, shown in Fig. 3.3. The 'percolation interceptor' was made a special feature: it consisted of units of no-fines concrete, drained through 100mm pvc pipes into the upper and lower galleries. Provision was made for passing a borehole TV camera through these pipes so that if any severe leakage should develop, its position could be determined. The dam has been described by Gadd et al. (1972).

The use of high cement-content concrete accentuated the problems of heat of hydration and shrinkage on cooling. The complexity of construction in alternate bays and watertight contraction joints increased the cost, often making an alternative embankment design a more attractive option, even when foundation conditions were suitable for a concrete dam.

The cost of concrete dams have been reduced by applying the construction methods used for embankment dams. By providing a separate waterproof element, the 178m-high Alpe Gira dam was designed with a relatively thin gravity section that would not be subject to uplift pressure or water pressure within its pores and so could be porous with open contraction joints. A low cement-content concrete, transported in 6m^3 dump trucks, was spread to 0.8m layers by bulldozer and consolidated by vibrators mounted on tracked vehicles. Open contraction joints were formed by cutting the layers with a hydraulically operated vibrating saw blade mounted on a self-propelled chassis. The dam was faced with 3mm thick sheet steel, backed by drains in case of any leakage. It was completed in 1964 and has been described by Gentile (1964) and Terracini (1970).

Further reduction of heat of hydration has enabled concrete to be placed by earth-moving machinery without contraction joints. An early use of rollcrete was described by Lowe (1962) and special mixes have been developed in preparation for a dam that was to have been built in Britain. Laboratory work and field trials were described by Dawson and Dunstan (1979). Elsewhere, several large dams have been constructed by this method. In Japan, the roller compacted dam (RCD) method of construction has aroused considerable interest. Shimajigawa dam (89m), completed in 1980, was the first such dam to be built and many more have followed. Yamauchi et al., (1985) have described the construction of the 100m-high Tamagawa dam, containing 1.14×10^6 m^3 of concrete, which started towards the end of 1983. By the end of 1998, according to Dunstan (1999), one of the highest Japanese RCD dams was the 155m-high Miyagase dam. In China, at the same time, the 131m-high Jiangya RCC dam was nearing completion. Construction of the 188m-high RCC Miel 1 dam in Colombia due to begin in 1999, while the Longtan dam in China was to have a final height of 217m and it was expected that construction of this RCC dam would resume in 2000.

Improvements in the properties of plastic sheeting offered the possibility that such material could be used to form a watertight element for dams. So wide did the usage become that the International Commission on Large Dams prepared Bulletian 78 (ICOLD, 1991) about this subject. It showed that there were 70 dams incorporating geomembranes throughout the world. This strategy was first used on a concrete dam, the Lago Baitone, in 1971; and first used on a masonry dam, the Lago Miller, in 1976. In general, it was used as a remedial measure to stop leakage caused by deterioration of older dams. To ensure adequate drainage behind the impervious membrane, special sheeting was soon developed that incorporated, typically, 2mm thick PVC sheet heat coupled to a non-woven, needle-punched polyester (PET) geotextile for drainage, as well as providing additional dimensional stability and protection against puncturing, so that the composite sheeting could be applied directly to the upstream face of the dam. Experience with this method of adding an impervious element to existing dams has been given by Scuero and Vaschetti (1997), who have described several different installations.

The 66m-high Piano Barbellino gravity dam was built between 1926 and 1931, high in the Italian Alps. Exposure to the very severe alpine conditions resulted in significant leakage through the facing. Despite various repairs, the deterioration continued, and in 1986/7, major repairs included application of waterproof sheeting to the upstream face. The selection of this remedial measure was based on both technical considerations and cost. The only access to the dam was by cable car so that weight and size were of major importance. The sheeting chosen was a 2 mm thick PVC geomembrane heat coupled to a non-woven, needle-punched polyester (PET) geotextile in order to give additional dimensional stability as well as protection against being punctured to allow of installation directly against the exiting upstream face of the dam, thereby avoiding extensive surface preparation. The sheets were 2.05m wide and were made to the lengths needed to extend from top to bottom of the dam face. They were then lowered into position from the crest, heat welded together to form a continuous watertight covering and attached to the dam by vertical metal strips, as indicated by Fig. 3.4. The sheets were sealed to the new bottom foundation concrete beam, with provision for draining the PET to the downstream part of the dam. The total area of the new composite geotextile was 6000m^2, and the work of installation took only 110 days. As part of the remedial work to the dam, it was extensively instrumented so that it could be continuously monitored. Prior to repair, leakage was 11.61/s, and this reduced to 0.671/s, of which only 0.12 l/s was confirmed as coming from the upstream face: a value that has remained constant since 1987.

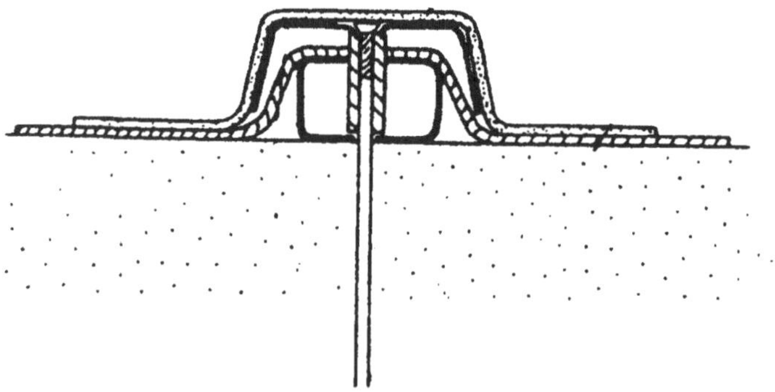

Fig. 3.4 Anchorage system for waterproof sheets

Amongst the numerous other examples of this type of repair is that of 178 m-high Alpe Gira dam described above, that was completed in 1964. Electrolytic action between the soil and the 3 mm-thick steel waterproof membrane, allowed leakage into the 0.3 m thick porous concrete drainage layer that was behind the membrane. The reservoir had been lowered several times during the years for maintenance of the steel plates, that had been treated with several times during the years for maintenance of the steel plates, that had been treated with vinyl acrylic paint, but only down to a certain level. Below this level, deterioration was becoming extensive, and it was decided to apply a PVC membrane over the whole of the lower part of the old steel membrane, covering an area of 11400 m^2.

Another example was the rehabilitation of the Cerresole 52 m-high gravity dam built towards the end of the 1920s (described above). This was an early example of the use of an upstream face formed of jack arches. Various repairs had been carried out to reduce the ensuring leakage, but it was finally decided to provide a separate upstream watertight covering. Various flexible liners such as mortar and resin coatings were considered, but in 1992, it was decided to use a PVC geocomposite, mechanically attached to the face of the dam to protect the face and avoid further leakage.

The value of the large volume of stored water in a reservoir can be very high, necessitating making repairs to the dam without lowering the reservoir level. Leakage problems with the Lost Creek dam in California were cured by the application of a PVC geomembrane underwater, with the aid of divers.

There are numerous other examples of the use of geomembranes applied to the upstream faces of dams to prevent leakage, protect a deteriorating face and stop the action of alkali aggregate cement reaction, which tends to cease in the absence of water. It has now reached the stage when the designs for new dams are being made with upstream membranes to form their impervious elements, so that the dam body can be made free draining and without risk of uplift pressures and unsightly staining on the exposed downstream surface. What could be regarded as a misconception in the design of gravity dams in the past has now apparently been corrected.

EMBANKMENT DAMS

Embankment dams of homogeneous section suffered the risk of instability caused by reservoir water reaching the downstream slope. Rotational slips occurred below the point where the phreatic surface meet the slope and along with clays, high pore pressures could cause deeper-seated slips.

The problem was solved by Terzaghi for the 34 m-high Vigario dam in Brazil, by separating the homogeneous section into waterproof and supporting parts with a vertical wall of filter material (Vargas 1970). This ensured that the water, under reservoir pressure, did not enter the downstream shoulder. The dam was re-named as the Terzaghi dam in his honour in 1964. Terzaghi applied this strategy to solve the problems at Sasamua dam. He replaced what had been designed as a central concrete core by a filter drain, as shown by Fig 3.5. At that time, designers were so used to providing zoned embankment dams with impervious cores, that this felt like turning the dam inside out, actually encouraging water to escape into the 'core' instead of being stopped by it. The work has been described by both Dixon et al., 1958 and by Terzaghi himself in an accompanying paper in 1958.

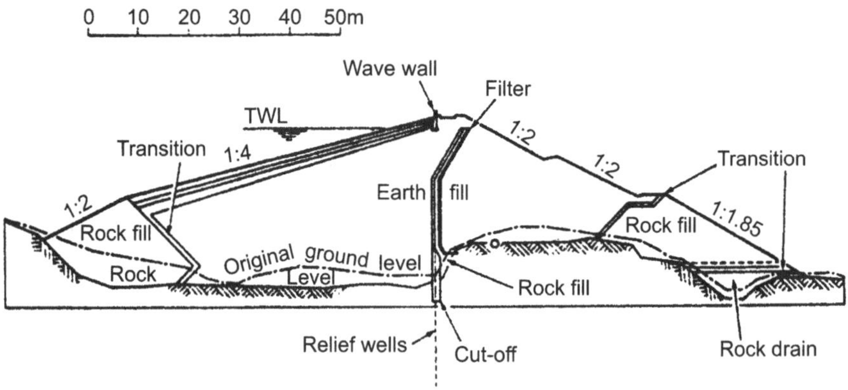

Fig. 3.5 Section of Sasamua dam

Clay Cores

The puddled clay waterproof elements of many old British dams are supported by shoulders of more permeable materials on either side. Often, a carefully selected fill was placed next to the core so that the permeability of the shoulder fill increased progressively with distance from the core. This provided drainage, often enhanced in the downstream shoulder by chimney drains built up from the formation drainage mattress at foundation level, dry stonework vertical columns to keep pace with the rising fill level, coming up into berms on the downstream slope where they also acted as surface drains. These are the true 'chimney drains', not to be confused wall drains that are often mistakenly called chimney drains.

The more modern central rolled clay cores are also supported by shoulders on either side, and drainage is provided by the use of filter drains built in contact with the faces of the core. These drains, in the form of continuous walls, must transmit the thrust from the core to the supporting structural shoulders and be so cleverly designed that water percolating from the core cannot carry clay particles away from the dam through the drainage system.

As we can see from the Maurice Levy approach, the total compressive stress across any potential plane through the core must exceed the pressure from the reservoir water in order to avoid any hydraulic fracture. Many designers make their clay cores curved in the plan so that if they are forced into a straighter shape by the thrust from the reservoir water, they will tighten up, increasing the total stress in the direction of the core axis. This may be a misconception, as can be seen in the following observations.

Measurements on the downstream faces of two clay cores made with horizontal plate gauges, reported by Penman and Charles (1973) and Penman (1973) show that during impounding, horizontal movements were negligible. This indicated the horizontal thrust from the clay cores on their shoulders exceeded the thrust from the reservoir water. Such thrusts caused some spreading of the embankment, so that the upstream tended to move upstream during dam construction, but the entire dam was not pushed downstream by the filling reservoir. Thirty years ago, construction on curved axis required difficult setting out, so most embankment dams were built with a straight axis. To check on the effect on price, the tenders for a dam in Cyprus were invited for both a straight and curved axis. The prices for a curved axis were applicably higher, reflecting this extra difficulty. The use of more modern recoding surveying instruments, combined with computer-made calculations, has removed this difficulty and during the design of the Megget dam in Scotland—the first in Britain to have a central asphaltic core—prices for a curved core attracted no greater price than if the core was to be on a straight axis. The dam was built with a curved core for aesthetic reasons, and the narrow core was set out from a stable observation pillar, built into the rock of the valley side at the left end of the dam, using a recording theodolite, precision mounted on the top of the pillar, observing a special staff held on the surface of the fill and moved to the required position with the aid of a radio telephone. Marks made on the fill, offset from the position for the core, were followed by the driver of the core-laying machine.

A clay core remains watertight because its self weight develops a sufficient amount of total pressure to hold the clay together and prevent it from being pushed apart (hydraulic fractured) by pressure from the reservoir water. The bulk density of core material is usually more than twice that of water, so the reduction factor from the overburden pressure σ_0 to σ_a can be 2, without any danger of hydraulic fracture. But clay in a vertical core is like material stored in a silo. During construction, the increasing weight produced by the increasing height causes compression and consolidation of the lower parts, and the clay moves down within the 'silo'. Adherence to the sides transfers some of the weight of the core to the shoulders, thereby reducing the vertical total pressure in the core material. The stronger the core material, the greater this reduction, so it can be seen that there is greater risk of hydraulic fracture in strong rather than in weak core materials. Puddled clay had to be soft enough for it to be effectively placed and compacted by the heels of the 'puddle gang'. It had a typically c^u value of 10 kN/m^2, ensuring an adequate total pressure in the clay if the core was not made too narrow. A misconception was the post puddle clay era of the use of drier and, hence, stronger clay for rolled clay cores. One reason for the use of the drier and harder clay was that current placing and compacting machinery had difficulty in working with clays to the extent of approaching the softness of the old puddle clays. The development of the wedge foot compactor enabled weaker clays to be used. In some ways, they resemble a caterpillar tractor that has lost its tracks. The four-toothed wheels that carried and drove the tracks are replaced by four broad rollers sprouting short, wedge-shaped steel 'feet'; counterpart of the 'sheeps' feet' of a sheeps' foot roller, but of much less length, more like blunt, wide, short chisels ends. Their tapered shape helps to throw off the sticky clay, cleared by scrapers passing between each row of wedge feet. The use of these machines has enabled the rolled clay cores to be made soft enough so as to be both flexible and less susceptible to the risks of hydraulic fracture.

Pore Pressures

An indication of total pressures may be obtained from measured pore pressures. In a saturated fill, any increase of total pressure will be carried initially by the pore water. When first placed, a rolled clay core can be expected to contain 3% to 5% air. This makes the pore fluid more compressible and despite the high compressibility of the structure of the clay, the ratio $\delta u/\delta\sigma = B < 1$ increasing total pressure as height increases causes the air to go into solution in the pore water and eventually increase the value of B to 1. If the clay was placed at a high enough water content to make it soft, then the value of B may soon rise to unity in the lower part of the core where any increase of pore pressure will correspond to the increase of total pressure. Comparison of these pressures with the calculated overburden pressure will give an indication of the loss of vertical total pressure caused by weight transfer to the shoulders on either side of the core. This action, as in silos, is referred to as arching action, because with sands in a silo, the sand can form arches across the width of the silo, developing bursting pressures on the silo walls; and because of the frictional properties of the sand, causing it to grip the sides of the silo and refuse to move down when an attempt is made to empty the silo. Beavan et al., (1977) have given several examples to prove that δu can be used as a measure of the increase of total vertical pressure $\delta\sigma_v$ in a clay core.

At the 36 m-high Derwent dam, a clay blanket was used to connect a central clay core to the upstream toe (this aspect will be discussed later). Pore pressures in the blanket and the lower part of the core, responded to values of total vertical stress that had been calculated with an allowance for an arching action across the with of the core, assuming that it adhered to the shoulders on either side, as indicated by Fig 3.6. During fill placement, pore pressures rose, as shown by Fig 3.7, show an excellent response, and dissipate by drainage during temporary cessations, such as the winter shut-down periods. The values plotted in Fig 3.6(1), as given by Rowe (1970) are the sum of the undrained increments and represent the pore pressure that would have developed if the fill had been placed instantaneously, leaving no time for drainage. These findings concur well with the calculated values, indicating that the measured values of pore pressure can give a very good indication of developing values of the total vertical stresses.

Comparison of the calculated total vertical stress and the measured pore pressures, Fig. 3.6 (2) and (1), with the overburden pressure (3) clearly shows the reduction of σ_v under the shoulders on either side of the core.

The distribution of maximum construction pore pressures observed at the 52m-high Kielder dam, described by Millmore and McNicol (1983), displayed a similar arching action of the clay core between the up and downstream shoulders, as shown in Fig. 3.8. Of particular interest are the high pore pressures measured in the clay blanket that had been placed under the upstream shoulder. Even in the core itself, the piezometric level was above the top water level at the end of construction. This was a design feature, following the argument made by Penman (1979) that the end of construction pore pressures should be at least equal to the pressure to come from the reservoir water, in order to ensure that the total pressures will be great enough to avoid hydraulic fracture.

In the past, high pore pressures in the shoulders of earthfill dams have led to instability. In order to avoid this phenomenon, placement water contents of shoulder fill have been reduced and it is now common practice to place drainage layers in the shoulder fill so as to minimise

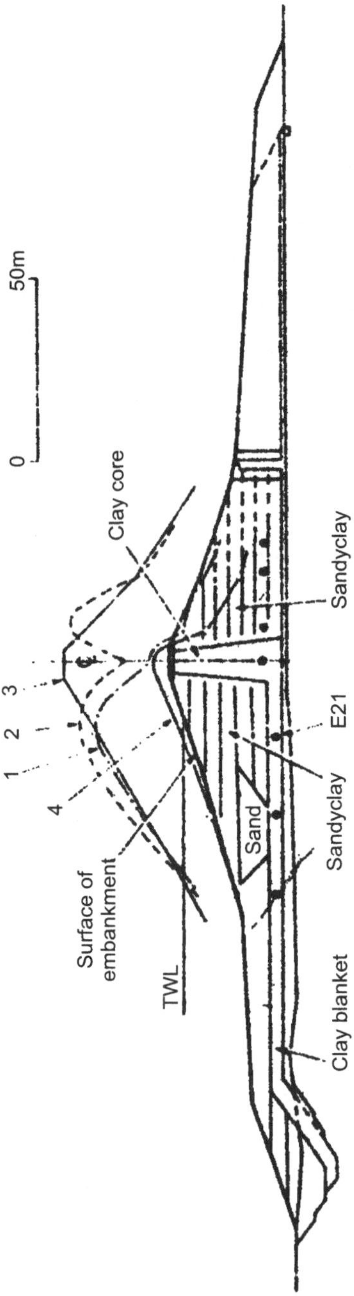

1. Sum of undrained increments of pore pressure
2. Calculated total vertical stress
3. Overburden pressure
4. Pore pressure at end of construction

Fig. 3.6 Derwent dam—distribution of pressures in base clay layer

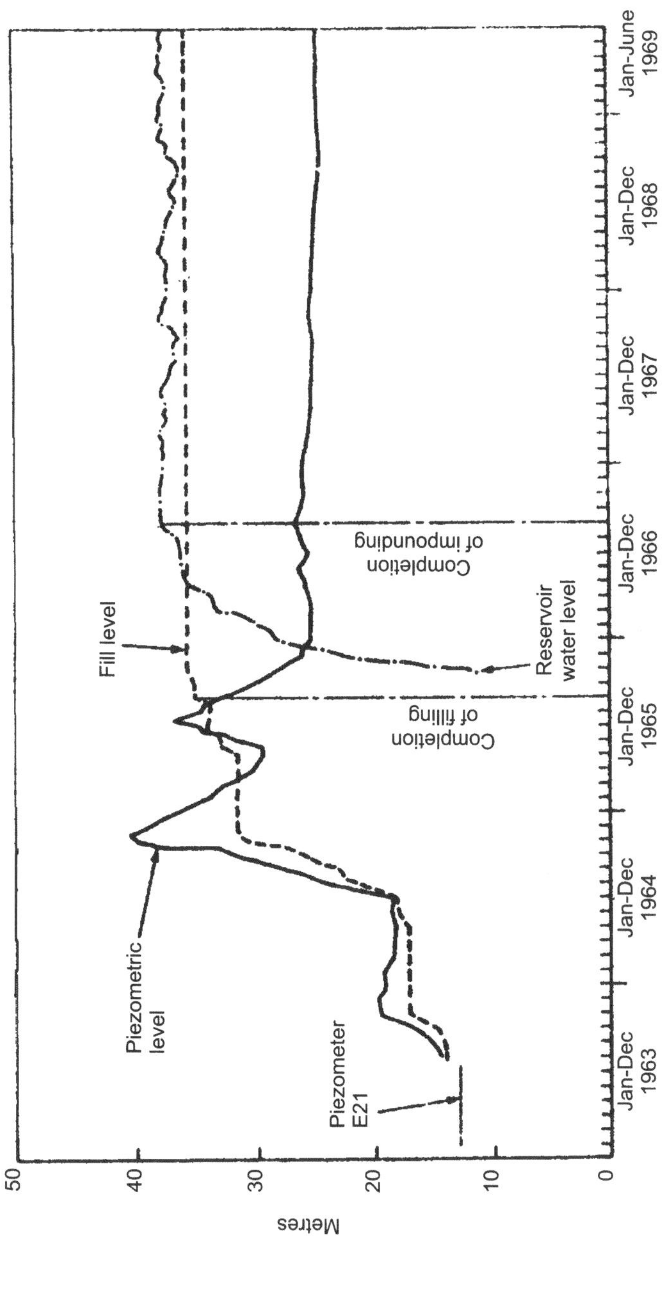

Fig. 3.7 Derwent dam—pore pressures measured by piezometer E21

construction pore pressures. Unfortunately, the concept of low placement water content has also been applied to rolled clay cores, and this must be regarded as a misconception.

It is normal practice to make the core width a function of depth from the crest, to maintain a constant hydraulic gradient over the whole height of the core. This is accomplished by sloping both sides of the core from crest width to a greater base width. At the crest, since it is above the top water level, core width could be zero, except that in practice there is a minimum width of core that can be constructed by machine. This width is usually added to the theoretical width required to give the desired hydraulic gradient, over the whole height of the core. In some cases, when there was a shortage of suitable clay, this minimum crest width has been kept constant down to a depth below the crest where the core width is equal to the required theoretical width, before the sides of the core were sloped outwards. Doing so produced vertical sides to the upper part of the core, which appears to inhibit settlement and may contribute to sufficient stress reductions to initiate hydraulic fracture. In two dams that experienced hydraulic fracture, cores with parallel upper parts had been used, although there were also other features that would have contributed to a reduction of the total stresses in the cores. It might be considered as a misconception to attempt an economy of clay volume by making the upper part of the core with vertical sides.

It should be pointed out that initial pore pressures in a rolled core cannot help but be below atmospheric pressure—i.e. it must be a suction—in order provide the effective stresses necessary to give the core material sufficient strength to support the placing and compacting machinery. It is usual to express pressure in terms of the ratio $r_u = u/\sigma_0$, and initial values of r_u will be negative, because $\sigma_v < \sigma_0$, r_u cannot be expected to rise to 1 in a core. The condition of pore pressure equal to reservoir water pressure will usually be satisfied with an end of construction value of $r_u \approx 0.5$ because the bulk density of compacted core material is commonly about twice that of water. It is clear from the above discussion that r_u values do not give a direct indication of σ_v.

The overall in situ permeability of clays is usually higher than that measured on samples in the laboratory because of the natural fabric of the material, as described in detail by Rowe (1972). The reason for puddling clay was to destroy the fabric and mix in any thin layers of sand or silt to make—if necessary with the addition of water—a soft uniform material of low permeability that could readily be pressed tightly into place under the pressure of men's feet, or more particularly, the heels of the puddle gang.

By contrast, clays for rolled cores often come straight from a borrow pit and may contain remnants of the original fabric. The action of the heavy placing machinery and rollers can form shear surface, e.g. due to bearing capacity failure under the heavy wheel loads; so they tend to have a mass permeability much greater than that of a puddled clay made from the same borrow pit. Because the design hydraulic gradient is usually lower than for a puddled clay core, this may be of no consequence, but the fissures and partings can form undesirable faults for the initiation of hydraulic fracture. It is, therefore, particularly important to ensure that adequate total stresses exist in a rolled core.

A wide core reduces the silo effect and increases the contact area with the foundation and abutments. In the extreme, the dam may approach a homogeneous section. Clearly, a compromise has to be reached in relation to the overall stability. As with other planes through the core, the total pressure across the contact must exceed reservoir water pressure to avoid leakage and risk of erosion. Arching across local irregularities may cause a dangerous reduction of the total pressure, and it is well known that the area of the abutments in contact with the core must be finished to a smooth surface without overhangs. This is often achieved with a concrete or gunite coating. Fissures in an otherwise competent bedrock can be particularly dangerous because the core material can form local arches across them and they provide a passage for water that can erode the core fill. Silt-sized particles can be washed through fissures less than 0.2 mm wide. Inspection of a cleaned abutment prior to core placement would normally consider such a fissure as closed. A concrete coating can isolate them from the core fill.

It is the usual practice to try to extend to waterproof element of a dam into weathered bedrock in the form of a grout curtain. If it forms a perfect barrier, it increases the length of the seepage path for reservoir water and its extent should be designed so as to reduce seepage flows to non-erodible values and bring the leakage within acceptable amounts. In many cases, piezometers placed upstream and downstream of a grout curtain reveal little head loss, showing that the permeability of the foundation has only been marginally reduced by the curtain. Casagrande (1961) made a special feature of this aspect and questioned the need for grout curtains in all circumstances.

At Sasmua dam, the original intention had been to form a grout curtain below the concrete core wall. Many holes had been drilled into the foundation when Terzaghi (1958) proposed the radical design change, as it were turning the dam inside out, by replacing the concrete core by a filter drain that, instead of stopping the water, would safely drain away any that got through the homogeneous upstream shoulder, thus ensuring that it did not pass into the downstream shoulder, thereby ensuring stability. Terzaghi formed the impression that the permeability of the jointed volcanic bedrock was not high enough to justify construction of a grout curtain, and replaced it by a filter drain using the existing grout holes which then behaved as relief wells.

The original Lugeon test, intended to determine the need for grouting, applied water to a borehole at a pressure of 1 N/mm^2. If the bulk density of the ground was 20 kN/m^3 and the assumption is made that the vertical total stress across a horizontal fissure is equal to then overburden pressure, then the application of 1 N/mm^2 pressure at the top of the borehole can be expected to cause hydraulic fracturing to a depth of 50 m. Grout applied under similar pressures would open up existing fissures and cause the ground to heave. Such disturbance of a weathered bedrock is undesirable.

Clearly, less disturbance will be caused by a given grout pressure if the vertical total pressure has been increased by the weight of the dam. There are cases where grouting has been carried out during winter shut-down by boring through the first season's fill, but maximum benefit could be obtained by waiting until dam construction is complete. This can be done by providing a gallery near the formation level that has been designed to give the headroom needed for the boring and grouting process. Galleries are often provided at the bases of upstream membranes and central asphaltic cores. It is also becoming quite usual to provide them at the base of rolled clay cores. Such a provision not only allows a grout curtain to be made after dam completion, but also allows the decision to be taken about the extent and density or even the actual need for a grout curtain at the later stage when the reservoir is filling, when it is possible to check on the development of any excessive leakage that could be stopped by a grout curtain. A great difficulty arises due to the financing of a dam. The designer has to estimate the total cost of the dam, and he is obliged to include money for a grout curtain. He dare not take the risk that a grout curtain will not be needed, because it—on reservoir filling—excessive leakage occurs under the dam, it is then too late to obtain more money for the installation of a curtain. Such a post-construction installation would be regarded as a failure of the original design, rather than a potential large saving of money.

Impervious Blankets

Towards the end of the nineteenth century, uplift and piping troubles with weirs on deep beds of sand in India were cured by the use of clay blankets placed upstream. The length of the blanket upstream was made sufficient to reduce the hydraulic gradient in the sand under the blanket/ weir combination to predetermined design values. These values, given by Bligh (1910), varied from 0.05 for silty sand to 0.1 for graval.

This approach has been used to reduce the hydraulic gradient under dams—as an alternative to a grout curtain—by extending a central clay core to form a clay blanket under the upstream shoulder. The effect on stability is clearly quite marked.

A central clay core designed to avoid hydraulic fracture will impose horizontal thrusts on the shoulders equal to the thrust from the reservoir water. The presence of a clay layer under the upstream shoulder will encourage horizontal sliding of the upstream shoulder.

The clay blanket placed under the upstream shoulder of the 36m-high Derwent dam was used because of difficulty in forming a central below-ground cut-off. At the site, bedrock lies 55m below ground surface at the deepest point. Deposits above the rock include a 14m-thick layer of laminated clay at a depth of about 13 m, underlain by a sand and gravel aquifer at an artesian pressure giving a had of 6 m above ground level. The layer was overlain by boulder clay about 4 m thick, covered by various silty clays and sands. The original intention was to construct a concrete wall in trench as cut-off down to bedrock and work began on its construction from the valley sides. The laminated clay was too soft to support the proposed dam and to avoid a deep-seated rotational slip under the shoulder, it had to be drained to be able to consolidate and gain in strength as the embankment was constructed. Vertical sand drains of 0.3m diameter were installed at spacing of 4.3 to 7.5 m under the entire embankment to pass through the layer into the aquifer.

Ground water lowering in the deep trench proved difficult and it became necessary to consider an alternative arrangement with the cut-off terminating in the boulder clay over the laminated clay. To prevent reservoir water passing under the cut-off via the sand drains and aquifer, it was necessary to place the impervious blanket over the upstream drains, i.e. under the upstream shoulder, to key 3m into the boulder clay at the upstream toe (Fig. 3.6). The existing sections of concrete cut-off in the valley sides were connected to the toe key trench.

Both rolled core and the 3.7m-thick clay blanket were placed about 3% above the optimum water content to give an undrained shear strength, c_u = 90 kN/m^2, in order to minimise the risk of cracking under differential settlement. The most critical slip surface passed through the clay core and blanket.

The fill was raised 12.8, 13.4 and 10 m to full height during three seasons. High pore pressures during the last season reduced the calculated factor of safety to an unacceptable value and fill placement was stopped for three weeks when the dam height was 2.7m below crest level in order to give time for dissipation of pore pressures. Horizontal movements at the upstream toe increased from about 3 mm to 13 mm during the last three months prior to completion.

At the 52 m-high Kielder dam, it was decided to avoid a major foundation cut-off and control underseepage by use of a clay blanket from the central core to key into glacial till at the upstream toe (Fig. 3.8). It was designed to reduce the hydraulic gradient under the dam to no more than 0.1. With the aim of developing construction pore pressures in the central core higher than the reservoir water level in order to avoid hydraulic fracture, a shear strength specification c_u = 60 to 140 kN/m^2 was used. As at Derwent, the critical slip surface passed through the clay blanket and core. Major placement was during two seasons when the fill was raised to 14 and 22.5 m. Towards the end of the second season, high pore pressures caused placement to be stopped prematurely. During the third (final) season, an increase in the pore pressure reduced the factor of safety to 1.28 with the fill more than 7 m below crest level. A small toe weighting berm was constructed to ensure stability as the construction was completed. During three months before the berm was built, horizontal movements of the upstream toe amounted to 100 mm as the dam was raised 7.5 m.

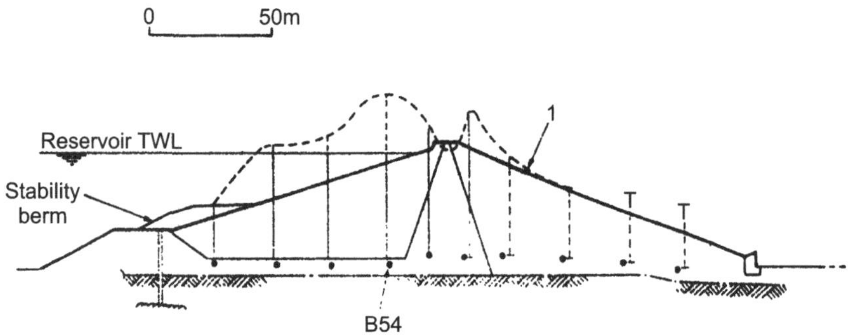

Fig. 3.8 Kielder dam—distribution of pore pressure in base layer.
1. Maximum pore pressures developed during construction. Values of B at piezometer B54 rose to 1.7

These two examples illustrate the care necessary to preserve stability when a soft clay blanket is used under the upstream shoulder. Cases where this feature has led to failure of the upstream shoulders have been discussed by Penman (1986) and will be described in other chapters of this book.

Progressive Failure

Stability analyses of slip surfaces commonly use the soil strength parameters c′ and ϕ′. These parameters are determined from shear tests on the soil and usually relate to maximum or peak measured strength.

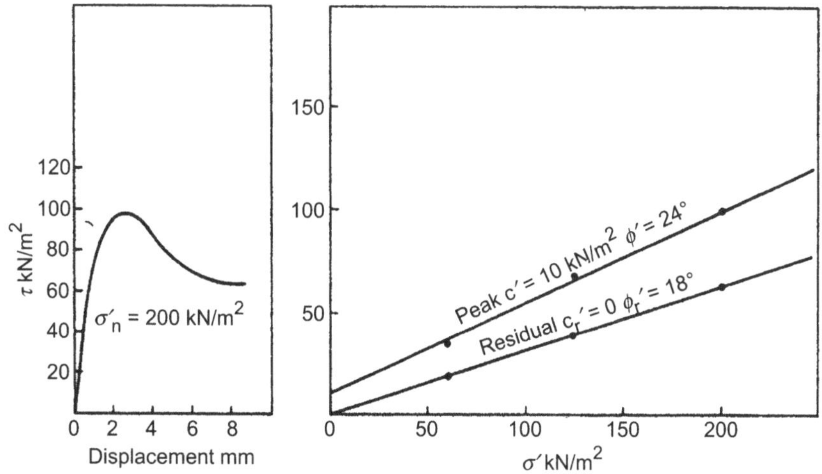

Fig. 3.9 Shear characteristics of a brittle soil

Many soils, when tested at a constant rate of strain, show a reduction of strength as strain is increased beyond peak, as indicated by Fig. 3.9. If the testing apparatus can apply sufficient strain, the strength will become constant at a residual value. These residual values give c'_r and ϕ'_r.

Traditional stability analyses (most of those adapted for calculation by computer) make the tacit assumption that strain along the entire length of the potential slip surface will be uniform. This may be regarded as a misconception. In practice, the strain along most slip surfaces is far from uniform. Bishop (1971) has suggested two types of variation that might occur along a simple circular surface. If the factor of safety used in the analysis is low enough, there is a danger that peak values may be exceeded on some parts of the slip surface. The subsequent strength reduction in brittle soils will throw extra shear stress on to other parts of the slip surface and increase the strain sufficiently enough to carry successive sections beyond peak. This aspect has been considered in detail by Skempton (1964 and 1984).

To avoid progressive failure, design must ensure that peak strength of brittle soils is not exceeded anywhere along the potential slip surface. If would clearly be safe but uneconomical to use ϕ'_r but a working compromise may be to use ϕ' and assume $c' = 0$. Progressive failure can occur in brittle soils when use has been made of c' and ϕ' in design with too low a factor of safety. An excellent analysis of the mechanism of progressive failure, following the failure of Carsington dam, was made by Potts et al., (1990) and further information about the method and examples of practical application to cut slopes was given by Potts et al. (1997).

Existing Shear Surfaces

A foundation soil may contain shear surfaces, formed at an earlier time, when the topography may have been different. Part of the surface of a rotational slip in a steep valley side may lie under the much gentler slope found in the valley today. Failures that occurred under the ice load may have left slip surfaces in areas where present topography gives no hint of past instability. Periglacial action can cause creep on relatively flat slopes, leaving the surface layers with large numbers of slip surfaces. Reactivation of shear on these surfaces can be expected to occur at residual values and design must assume ϕ'_r and c'_r (probably zero).

It is extremely difficult to discover these pre-formed slip surface during site investigation. It is also almost impossible to detect these surfaces from boreholes and considerable experience is required before they can be identified in trial pits. A knowledge of the geological history of the site may offer the best indication of the likelihood of slip surface and their identification may be assisted by some slight change in the properties or quality of the soil above and below them. If the geology of the site gives any suggestion that there may be shear surfaces along which a slip might pass, the design assumption should be made that they do exist unless their absence can be proved by a site investigations.

In design, the stability analyses used must consider slip surfaces that will really pass through the potentially weakest layers. The speed of computer analyses enables large numbers of surface shapes to be checked, but there is always the danger that the programme being used does not allow for all of the worst conditions.

ROCKFILL

At one time, specifications for dumped rockfill required the use only hard, competent rock from the heart of the quarry and inclusion of no small sizes. Weathered rock was wasted and often quarry run was passed over a grizzly in order to remove fines. It is now a recognized fact that this strategy was a misconception.

Contact forces in large, single-size rockfill can be alarmingly high. They are approximately proportional to d^2 (where d represents particle diameter). In fill consisting of 0.5 m dia. pieces, an overburden of 100 m could produce a vertical force of about 400 kN on each contact, whereas a single size coarse sand with 0.5 mm dia. grains, would have to carry only 0.4 N at each contact.

Hard rock (granite, basalt) usually quarries to give very angular shapes. The sharp points of contact are crushed under high loads until the contact area is sufficiently increased so that the strength of the parent rock can provide the necessary support. The effect of water is to reduce the rock strength, so that further crushing will occur when the rockfill is wetted. The open voids between large pieces of rockfill allow them to rotate when further contact crushing occurs, which may result in large overall settlements of the rockfill.

The remedy is to support each individual large piece of rock on a bed of fines, akin to placing masonry blocks on mortar. Use of sufficient fines (down to sand size) not only avoids excessive contact pressures between the larger pieces of rock, but also completely fills the voids and prevents rotation. Compaction at optimum water content by heavy vibrating smooth rollers can produce a very satisfactory fill. No test has yet been devised to determine the optimum water content for large sized rockfill, but fortunately, a slight excess of water is unlikely to be damaging because it can safely drain away. Penman and Charles (1976) have suggested that the limit to the size and amount of fines is when the in situ permeability of the rockfill is reduced to 1×10^{-5} m/sec. Material with an in situ permeability much below this value may become subject to construction pore pressures and should be regarded as earthfill.

DAMS WITH UPSTREAM MEMBRANES

If the horizontal thrust from a central clay core on the up and downstream shoulders is equal to the thrust from the reservoir water, then one shoulder should provide sufficient restraint to retain the reservoir. Use of an upstream membrane transfers the water pressure as a thrust onto the fill which—if it can have sufficiently steep slopes—need be of no greater volume than one shoulder of a central core dam.

The 84 m-high Shing Mun dam (mentioned above) is a unique example of a type of rockfill dam built with a section similar to that of a downstream shoulder (Fig. 3.2). Thrust from the reservoir water on the waterproof slab (4) was transmitted through the buttresses faced with pre-cast blocks (8) to a wide concrete thrust block (5), forming the upstream face of hand-placed rockfill (6). A sand cushion (7) was used between the block and rockfill to avoid stress concentrations. The wide drainage gap between the separate waterproof slab and the thrust block ensured that no reservoir water reached the supporting rockfill. This hand-placed rockfill could be expected to be much stiffer than the dumped rockfill, commonly used for upstream membrane dams at that particular time.

Dumped rockfill earned the early upstream membrane dams a poor reputation because the large deformations that commonly occurred on the very first impounding often caused severe damage to the membrane. British practice did not favour them. The first large rockfill dam with an upstream membrane for an impounding reservoir in Britain was not built until 1975. The vastly improved behaviour of compacted rockfill has brought them into fashion throughout the world. The highest for many years was the 160 m Foz do Areia in Brazil, but even higher ones are currently under construction, and many more are planned.

The weakness in rockfill lies in the high hydraulic gradient across what is almost a line contact between membrane and foundation or abutment. It is usual to place a grout curtain along this line, sometimes with a gallery incorporated in the grout cap so that further grouting can be carried out if found necessary. It is difficult to design this type of structure without requiring some fill on its downstream side and the compression of this fill, under full reservoir loading, can cause a sharp differential movement in the membrane as it passes from the well founded, rigid concrete structure onto the fill.

The situation may become much worse if it is necessary to increase the height of the toe structure. This can occur when any design based on a site investigation—has decided on a formation level after removal of top soil, slope wash and severely weathered material. Construction logistics may require construction of the grout curtain to be amongst the first works on the site, thereby fixing the line for the toe structure. If ground conditions are subsequently found to be worse than assumed and formation level has to be lowered, the toe structure has to be made higher, resulting in a greater depth of fill immediately adjacent to it.

On sites where formation level can be formed in a competent bedrock, the toe structure can be reduced to the form of a slab, e.g. (Fig. 3.10) Deer Creek (85 m), Salvajina (148 m), thereby minimising fill thickness to virtually zero next to the structure. At Khao Leam (130 m)—because it was founded on karst, it was felt desirable to have a gallery—this was built above the slab to avoid any backfilling that would have to carry the membrane, i.e. to eliminate any risk of differential settlement.

Often this problem only occurs near the streambed, but this is where reservoir head is at the maximum and the effects of differential settlements may be most acute.

INSTRUMENTATION

Instrumentation is required to see whether design conceptions approach reality. Type and positions for measurement can only be decided by the designer, who knows the weaknesses and peculiarities of the site and dam. There is a danger that choice and positions for instruments is not

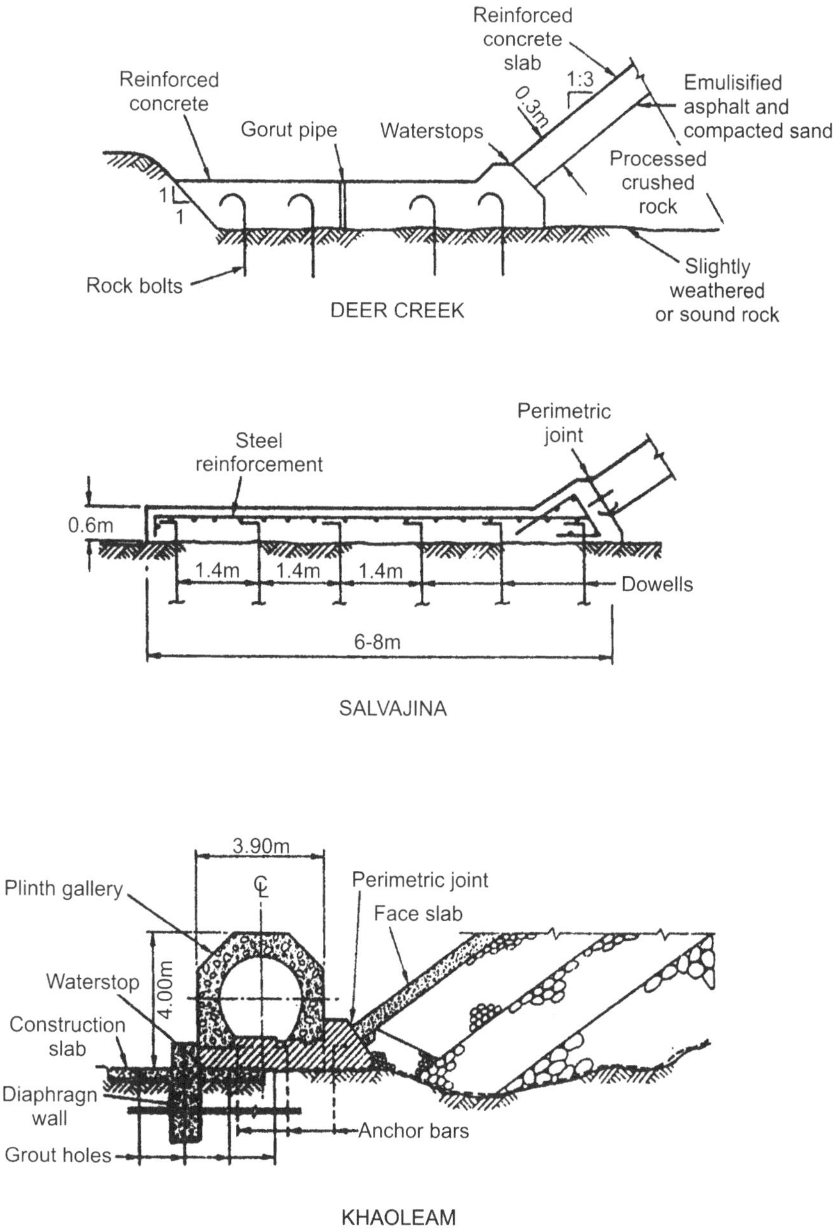

Fig. 3.10 Toe slabs of three large upstream membrane dams

given the detailed consideration it deserves. There are cases where instruments have been positioned in neat rows at several levels across a few cross-sections, without reference to potential weaknesses nor full appreciation of the resulting workload of taking and interpreting readings. The cost of instrumentation lies largely in installation and subsequent reading and analysis of the measurements. The cost of the instruments themselves is relatively small and it is always worth using the very best types in the expectation of reliability, accuracy and longevity.

It is becoming the norm for instruments to be supplied and installed by specialist firms as subcontractors to the main contractor. They are thus isolated from the designer and, due to cash flow considerations, often not engaged until a late stage, so that production and supply of some instruments may have to be completed in an unnecessary short time to suit construction schedule. This does not encourage full trails and repeat calibrations.

As with site investigation, competitive tendering forces the prices to come down and can lead to less than satisfactory results. The vital work of installation may be entrusted to a minimum number of poorly paid staff. Often the required instrument chambers or gauge houses have not been built and the sub-contractor's staff lack the authority to address the problems to the designer or owner. This may cause omission of the initial and repeated readings so necessary to establish zero conditions. Without reliable zero measurements, subsequent readings may be of limited value.

An apparently simple measurement of considerable importance is that of leakage. The permeability of the waterproof element and the surrounding natural materials ensures a minimum seepage that can be expected to vary with reservoir water level. Increasing flows can give early warning of deteriorating conditions.

Measurements that are commonly made of the flows from drainage systems of the dam may not provide a complete picture. They are often complicated by the effects of rainfall but, more dangerously, they may fail to record leakage escaping through permeable bedrock below the drains. More consideration should be given at the design stage to devising drainage systems that will collect seepage without loss into the downstream foundation, so that reliance can be placed on the measured discharge as a true record of leakage.

The rise of water level in the reservoir during impoundment raises the phreatic surface in the valley sides. Before the dam was built, and downstream of it, the phreatic surface within the soil and rock of the valley sides exits at the river level. With a full reservoir, the phreatic surface upstream of the dam exits at reservoir water level, but downstream of the dam it should exit at the river level. It may be difficult for the phreatic surface in both valley sides to fall from reservoir level to river level over the upstream to downstream width of the dam so that, with an embankment dam, the phreatic surface will exit into the dam fill and water from the ground will flow into the drains and may be measured as leakage through the dam, varying with reservoir level, exactly as a real leak would. This apparent leakage measured by flows from the under-drains can be considered as a misconception.

CONCLUDING REMARKS

At the design stage, the type of dam chosen for a new site is most often the embankment type because of its low cost and ability to accept foundation conditions unsuitable for a concrete dam.

Developments in the art and science of geotechnical engineering—combined with research studies of the behaviour of large embankment dams that have been carried out throughout the world—have given the designer reliable methods for designing safe slopes. Erosion by seepage and leakage leading to piping failure can be a serious danger and more attention should be paid at the design stage in order to eliminate this danger.

Perhaps one of the most dangerous design misconceptions is that the factor of safety against dam failure can be obtained from computer analysis of slope stability. There is no known method for calculating a factor of safety against piping failure.

Leakage through a core that causes erosion and the development of a pipe can be always said to originate in hydraulic fracture. If the total pressure across the potential leakage path exceeds the reservoir water pressure, the water can only percolate through the voids of the soil and leakage rates can be calculated from soil permeability. Excessive hydraulic gradients can cause loss of material at the discharge face, but this can be protected by filters. The dangerous situation arises when total pressures at any position are not sufficient to avoid hydraulic fracture and high water velocities occur, sufficient to erode the soil.

Sherard (1985) believes that 'there is now essentially incontrovertible evidence available to the profession supporting the conclusion that small concentrated leaks by hydraulic fracturing develop commonly in well-designed and constructed dams'. Fortunately, few lead to failure: careful observation allows remedial measures to be taken in time. Sherard refers to many examples of low homogeneous dams that failed rapidly by breaching on the very first reservoir filling. He also discusses several cases of large dams, the highest of which to fail was the 93m-high Teton dam in 1976. The discovery of 'wet' seams in the cores of several dams adds evidence to the occurrence of hydraulic fracture. A particular hydraulic fissure found passing through the wide silt core of Teton during post-failure investigations was particularly intriguing and led to speculation as to whether it could have resulted in failure, had the dam not first failed due to hydraulic fracture in the badly designed key-trench. Sherard has since (1987) expressed the view that the wet seam would not have caused failure because the horizontal fracture formed by the hydraulic separation pressure was closed by loose silt falling from the soffit as the opening progressed in the downstream direction, so that by the time it reached the downstream face of the core, the loose silt provided sufficient resistance to prevent a rate of water flow that could be erosive.

First indicators of impending trouble may be shown by changes in the rate of seepage through the dam. It is of particular importance to be able to accurately measure the flow of all water passing through a dam. Although measurement is extremely simple and can be easily and continuously recorded, it is usually found to be almost impossible to collect all of the seepage water and difficult to separate the effects of rainfall. Much more consideration should be given to this aspect at the design stage itself.

REFERENCES

Beavan GCG, Colback PSB and Hodgson RLP (1977). Construction pore pressures in clay core dams. *Proc. 9th Int. Conf. Soil Mechanics and Foundation Engineering,* Tokyo, vol. 1, pp 391–394.

Binnie WJE (1936). Study of the facing of masonry and concrete dams. *Trans. 2nd Int. Congress on Large Dams,* Washington, vol. 3, pp 343–355.

Bishop AW (1971). The influence of progressive failure on the choice of the method of stability analysis. *Geotechnique,* Vol. 21, No 2, pp 168–172.

Bligh WG (1910). Dams, barrages and weirs on porous foundations. Eng. News Record, Vol. 64, pp 708–710.

Casagrande A (1961). 1st Rankine Lecture: Control of seepage through foundations and abutments of dams. *Geotechnique,* vol. 11, no 3, pp 161–181.

CHINCOLD (1979). Dam Construction by the Chinese People. *The Chinese National Committee on Large Dams.* 237 pp.

Dawson ES and Dunstan MRH (1979). Trials for a prototype rolled concreted dam. *Trans. 13th Int. Congress on Large Dams,* New Delhi, vol. 9, pp 647–660.

Dixon HH, Edlington GA and FitzGerald EP (1958). The Chania-Sasumua water supply for Nairobi. *Proc. Instn. Civil Engineers,* London, vol. 9, pp 345–368.

Dunstan MRH (1999). Recent developments in RCC dams. *Int. Jnl on Hydropower and Dams,* vol. 6, Issue one, pp 40–45.

Gadd PJ, Mander GC and White TES (1972). The Wakefield and District Water Board's Booth Wood dam. *Water and Water Engineering,* vol. 76, pp 81–84.

Gentile G (1964). Study, preparation and placement of low cement concrete, with special regard to its use in solid gravity dams. *Trans, 8th Int. Congress on Large Dams,* Edinburgh, Vol. 3, pp 259–277.

Gröner CF (1933). *Method for repair and preservation of dam as used at the Ringedals dam.* A/S Tyssefaldene, Hardaner, Norway.

ICOLD (1974). Lessons from dam incidents. *Int. Commission on Large Dams,* Paris. 1068 pps.

Levy M (1895). Memoir submitted to Academie des Sciences, 5th August.

Lowe J (1962). Use of rollcrete in earth dams. Contribution to discussion. First ASCE Water Resources Engineering Conf, Omaha, Nebraska. (See *Proc. Int. Conf. Rolled Concrete for Dams,* CIRIA, London, 1981, pp W-1 to W-5).

Millmore JP and McNicol R (1983). Geotechnical aspects of the Kielder dam. *Proc. Instn. Civil Engrs,* London, Part 1, vol. 74, pp 805–836.

Penman ADM (1973). Contribution to discussion. *Trans. 11th Int. Congress on Large Dams,* Madrid, vol. 5, pp 520–522.

Penman ADM (1979). Interface problems of dams-contribution to discussion. *Trans. 13th Int. Congress on Large Dams,* New Delhi, vol. 5, pp 249–253.

Penman ADM (1982). Materials and construction methods for embankment dams and cofferdams. General Report. *Trans. 14th Int. Congress on Large Dams,* Rio de Janeiro, vol. 4, pp 1103–1228.

Penman ADM (1986). 26th Rankine Lecture: On the embankment dam. *Géotechnique,* Vol. 36, No 3, pp 301–348.

Penman ADM and Charles JA (1973). Effect of the position of the core on the behaviour of two rockfill dams. *Trans. 11th Int. Congress on Large Dams,* Madrid, vol. 3, pp 315–339.

Penman ADM and Charles JA (1976). The quality and suitability of rockfill used in dam construction. *Trans. 12th Int. Congress on Large Dams,* Mexico City, vol. 1, pp 533–556.

Potts DM, Dounias GT and Vaughan PR (1990). Finite element analysis of progressive failure of Carsington embankment. *Géotechnique,* Vol. 40, No 1, pp 79–101.

Potts DM, Kovacevic N and Vaughan PR (1997). Delayed collapse of cut slopes in stiff clay. *Géotechnique,* vol. 47, No 5, pp 953–982.

Rowe PW (1970). Derwent dam-embankment stability and displacements. *Proc. Instn. Civil Engrs,* London, Vol. 45, pp 423–452.

Rowe PW (1972). 12th Rankine Lecture: The relevance of soil fabric to site investigation practice. *Géotechnique,* vol. 22, No 2, pp 195–300.

Scuero AM and Vaschetti GL (1997). Italy's role in developing geosynthetics for hydraulic works. *Intn. Jnl Hydropower and Dams,* vol. 4, Issue 2, pp 86–93.

Scuero AM and Vaschetti GL (1998). A drained synthetic geomembrane system for rehabilitation and construction of dams. The Prospect for Reservoirs in the 21st Century. Proc. 10th BDS Conference. Thomas Telford Ltd, London, pp 359–372.

Sherard JL (1985). Hydraulic fracturing in embankment dams. *Proc. Symp. Seepage and Leakage from Dams and Reservoirs,* ASCE Spring Convention, Denver, pp 115–141.

Sherard JL (1987). Teton dam failure: Retrospective comments. Dam Failures, GA Leonards Ed. Elsevier Science Publishers, Amsterdam, pp 283–293.

Skempton AW (1964). 4th Rankine Lecture: Long-term stability of clay slopes. *Géotechnique,* vol. 14, No 2, pp 77–101.

Skempton AW (1984). Special Lecture: Residual strength of clay in landslides, folded strata and the laboratory. *Géotechnique,* vol. 35, no 1, pp 3–18.

Terracini C (1970). Behaviour of Alpe Gera gravity dam in its early operational stages. Trans. 10th Int. Congress on Large Dams, Montreal, vol. 4, pp 119–137.

Terzaghi K (1958). Design and performance of the sasumua dam. *Proc. Instn Civil Engrs,* London, vol. 9, pp 369–394.

Vargas M (1970). The use of vertical core drains in Brazillian dams. *Trans. 10th Int. Congress on Large Dams,* Montreal, vol. 1, pp 599–608.

Wegmann E (1927). *The Design and Construction of Dams.* John Wiley, New York, 740 pp.

Yamauchi T, Harada J, Okada T and Shimada S (1985). Construction of Tamagawa dam by the RCD method. *Trans. 15th Int. Congress on Large Dams,* Lausanne, vol. 2, pp 89–115.

Chapter **4**

Classic Distress and Failures of Dams

INTRODUCTION

Distress and failure in dams are not an uncommon occurrence and prove or disprove the abstract theories, when applied in practice or present an opportunity to develop theories based on practical observations. A crack or settlement in a dam, an intact dam along with a failed reservoir or the failure of a dam causing storage loss—these situations present difficult and sometimes confusing situations. In one case, the site is found suitable for reservoir and not for dam and at some other site, the reverse is the case. Whereas floods are automatically covered in consideration of hydrology of the site and hydraulic standards of the dam, earthquake faults simply make the site unsuitable for dam. When overlooked or underestimated, any erosion of faults causing displacements in structure have resulted in catastrophic failure. Leakage through an impervious membrane—resulting in piping or choking of drains—in case there is no alternative to take care of the situation, create an uncontrollable phenomenon which may result in the failure of the dam or reservoir. Instrumentation of dams is strongly advocated, but this does not guarantee against any distress or failure of the dam, unless instrument placement is proper or data collected is analysed in time to avail of a warning, if any.

There may be numerous other reasons of failure other than the ones mentioned and illustrated by the few case histories included in this chapter. There may not be just one reason but a combined effect of many. First the filling, followed by the rate of construction or progressive movements spread over a long period may result in distress or disaster. Investigations by all means do not reveal the exact as also all the happenings, resulting in controversial conclusions. Experiences of individuals play an important role in attribution of causes as, deduced by the post-failure observations and test data, revelation from site or evaluation of design and its suitability.

Worst is the situation when engineering judgment or technical reasoning is influenced by outside pressure, such as 'bureaucracy, politics and local lobbies'.

The presentation of the case histories in this chapter or elsewhere may be incomplete as their documentation in printed literature, by different authors and investigators also differ in data and observation. In the cases, re-examined after decades, when individual's interest is expected to be relatively un-biased and is guided by mere technicalities of the case, it is found that consensus could not be reached on reasons of distress' leading to failure and questions lurking in engineering minds, remain unanswered'.

TETON DAM

Teton Dam, the major feature of the multipurpose Teton Basin Project of the United States Bureau of Reclamation, failed on 5 June 1976 while the reservoir level was about 1m below the spillway sill. The dam was a central-core zoned earthfill structure, with a height of 93 m above the river bed and 123.4 m above the lowest foundation. The core material was wind-deposited nonplastic to slightly plastic clayey silt. A steep-sided key trench filled with the same core material but partly specially compacted, with a grout curtain built through a concrete grout cap in the trench base were relied upon for principal seepage control. The volcanic rocks at the damsite were highly permeable and moderately to intensely jointed. The wider foundation joints were treated for sealing, while the narrower onces were not treated.

The investigations following the failure of the dam have been discussed in the following pages.

The Failure Sequence

When a reservoir begins to fill for the first time, it is usual to check for any leaks from the new dam by looking into the stream bed and the valley sides downstream. As the filling reservoir formed by Teton dam reached half way and more, careful inspections were made, but no signs appeared until the water level had reached within a few metres of the spillway ogee. The ogee level was the lowest level where water could overflow from the reservoir, and in normal use, with the gates closed, the water level would be much higher, with design top water near the top of the gates. When the water level was about 3½ m below ogee level, some small springs were found at the riverbed level, about 45 m from the downstream toe of the dam. The following day, Friday, 4 June 1976, when the level had risen to 2 m below ogee level, some further springs were found about 12 m downstream. In total, these springs were only discharging about 40 l/s of clear water and caused no concern.

By Saturday, 5 June, the water level had reached within 0.9 m of the ogee level and when the first of the workforce arrived on site at 7 o'clock, they found dirty water coming out of the rock debris in the groin between the right valley side and the downstream slope of the dam near the dam toe at a rate estimated as 700 l/s. There was also a discharge of about 55 l/s coming from the slope of the dam, close to the same groin at about 2/3 full height of the dam. During the next three hours, this flow increased to more than 420 l/s; then only half an hour later, there was a loud burst and the point of issue moved to the left so that the water was coming from a place in the downstream slope, 5 m from the groin, as seen in Fig. 4.1. After that, the seepage increased rapidly.

Fig. 4.1 Water issuing from downstream slope at about 2/3 dam height and 5m from groin with the right abutment.

It was clear that conditions were deteriorating from an early stage, and at about 9 o' clock, a call was made to the Resident Engineer, who on this Saturday, with only routine work at the dam, was thinking of staying at home. When he arrived on site, he could see the danger of the dam falling,

Fig. 4.2 Rapidly increasing size of hole able to swallow bulldozers.

and got in touch with the police, who were able to issue a warning to all those living downstream, to move to higher ground. By 11.20, the eroded hole in the dam, shown in Fig. 4.2, was so large that bulldozers that had been attempting to stem the flow by filling the hole, lost ground so rapidly that they sank into the flow. The flow increased and the erosion continued rapidly, as seen in Figures 4.3, 4.4 and 4.5, until the dam crest breached at five minutes before noon and complete failure ensued. The flood flowing from the dam is shown in Fig. 4.6, and the discharge from the reservoir is shown in Fig. 4.7, where the wet upstream slope shows the amount of drawdown that had occurred in a few minutes. Because of the early warning and that all these occurrences were during daylight, the resulting flood caused the loss of only 14 people and of these, one was shot by his own gun as he rushed to climb over a gate and somehow caught the trigger.

Site Selection

The 93 m-embankment dam was the highest dam to have failed. It was of a fairly traditional design—seen by the cross-section in Fig. 4.8—with a

Fig. 4.3 Condition when a further swallow hole appeared and the whirlpool can be seen in the reservoir.

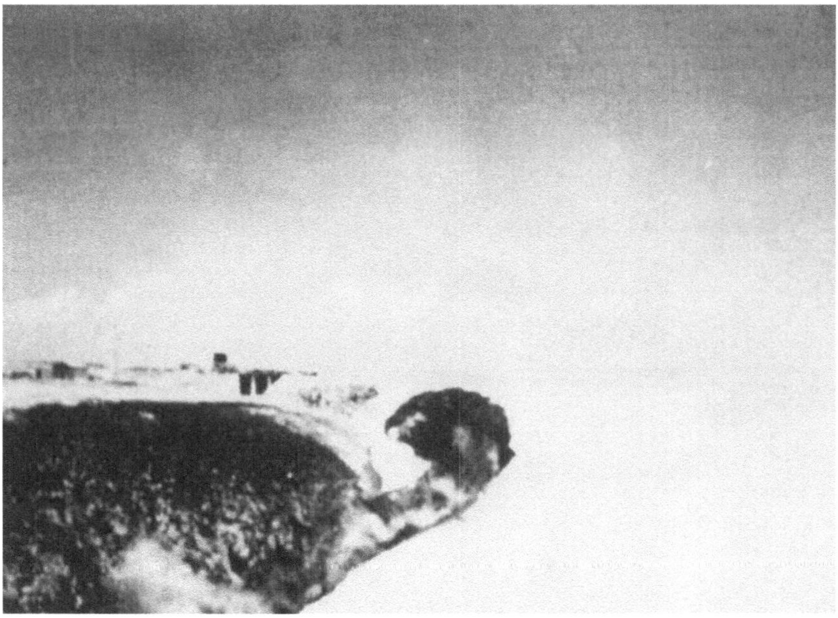

Fig. 4.4 View of Teton dam after failure.

Fig. 4.5 The dam has virtually breached, although the crest is still intact. Its collapse was timed as 11.55 when the power line was broken.

Fig. 4.6 Flood flowing from the dam.

Fig. 4.7 Discharge from the reservoir.

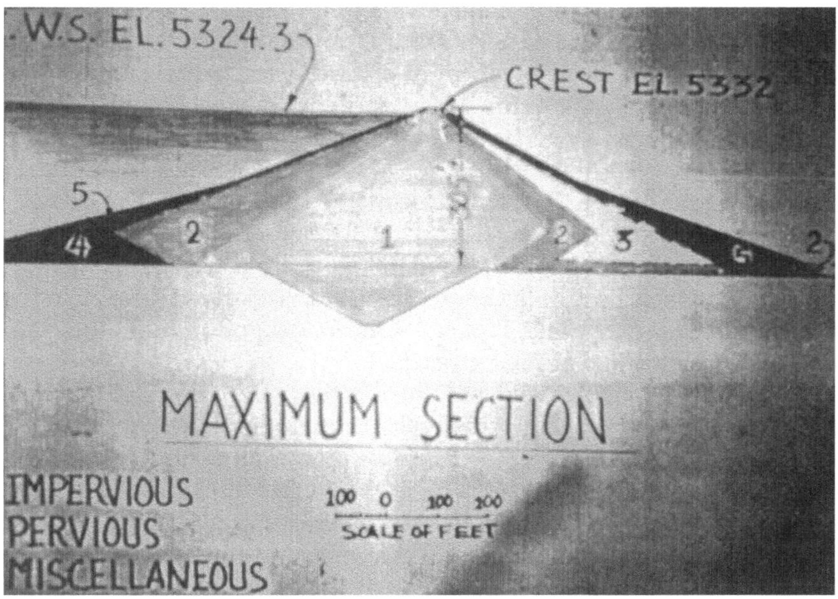

Fig. 4.8 Maximum section. (1) silt core. (2) selected sand, gravel and cobbles. (3) miscellaneous. (4) selected silt, sand, gravel and cobbles. (5) rockfill.

very wide central core, taken down in wide trench to the bedrock. The core was supported by small shoulders of sand and gravel with some miscellaneous fill and a small amount of rockfill on the downstream side. The site for the dam was most unsuitable and the dam was only built after the application of considerable pressure from the farmers of eastern Idaho, who had enjoyed the facility of an abundant supply of water from wells in the very permeable mass of volcanic rocks under their land. This land is conveniently covered by a thick layer of silt, rich in minerals, that will grow a good crop, given the water for irrigation. A well was reported to be producing 140 l/s with a draw-down of only 0.6 m. The volcanic rocks contain thick deposits of lake sediments. These rocks are much less permeable than the general mass of volcanic rock and well drillers were prepared to abandon wells found to be going into these sediments. The regional water table was 60 to 150 m below ground level and downstream of the dam site, the Teton River loses water into the volcanic bedrock. Under the reservoir area, sediments have created a perched water table and one of the factors that led to the reservoir area being considered suitable was the fact that the river was not losing water into the rock over this particular stretch.

It is clearly uneconomic to allow water to fall to the deep water table and then have to pump it back up again for irrigation and the farmers had been agitating for more canalled water for their land. A *raison d'être* for the Bureau of Reclamation was the provision of irrigation water. Consideration had been given by the Bureau to the development of water resources on the Teton River as early as 1904. More detailed investigations began in 1932: most of the sites studied were on the upper Teton River and its tributaries. The positions of several sites are shown in Fig. 4.9, and the two on Canyon Creek, marked as 6 and 7 on Fig. 4.9 were studied in 1946 and rejected. A site downstream of the present site was proposed for a 14 m high diversion dam to put water into a canal to take water to the North Fork of the Snake River, but a report in 1957 concluded that the site was unfeasible. Figure 4.9 shows the positions of other sites that were investigated and later rejected.

In 1957, two boreholes were put down at the present site. One borehole in the river channel showed 30 m depth of alluvium below riverbed, lying over the rhyolite bedrock, while the other on the left abutment, revealed rhyolite for its entire depth of 87 m. Representation from the Teton Country Wheatgrowers Association to the Bureau in 1961 suggested that they should consider building a reservoir in the upper reaches of the river. It could be concluded that pressure for the provision of a reservoir was intensifying. No doubt the costs of pumping irrigation water were considered, ever more closely as farming developed business efficiency. In November 1961, a Bureau geologist looked at five possible sites on the

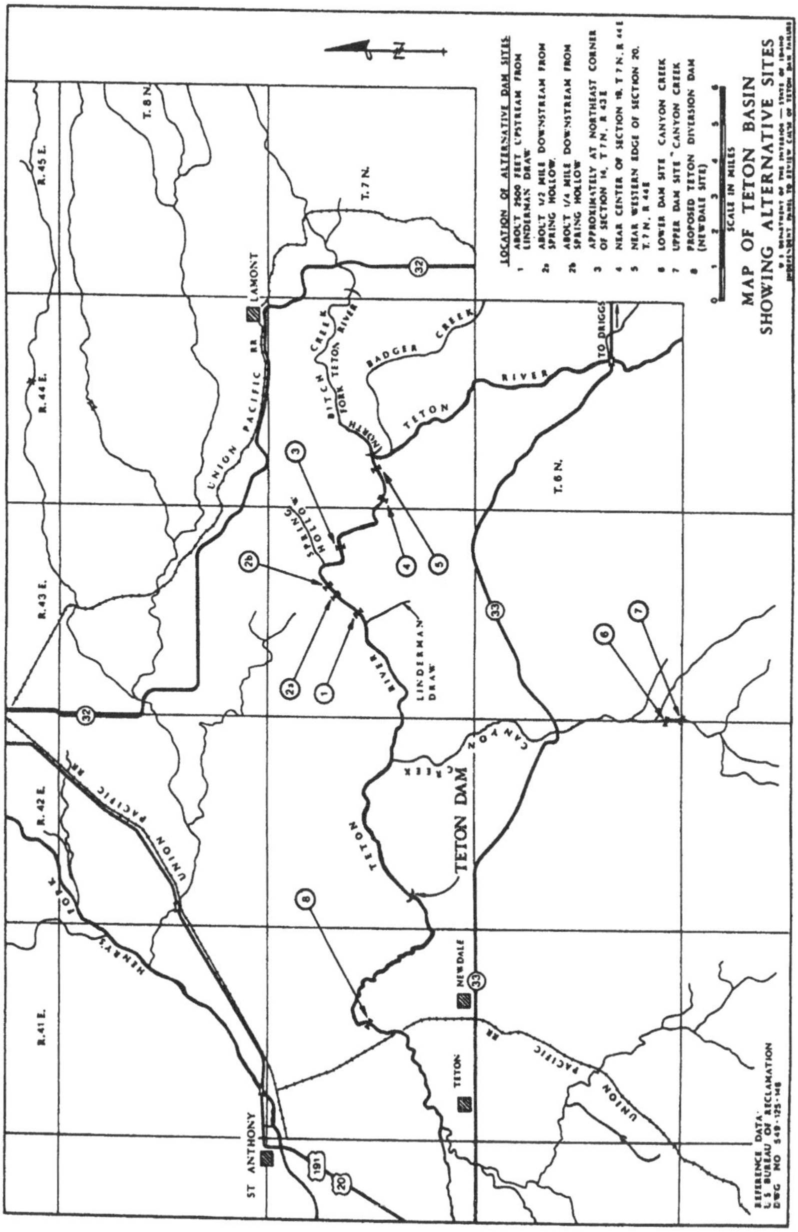

Fig. 4.9 Map of Teton basin showing alternative sites.

Teton river and early in 1962, a Bureau report recommended the present site as providing a suitable position for a large reservoir that would allow efficient supply of water to the required areas, be large enough for flood control and be over a stretch of river where the water was not lost to the underlying previous bedrock. The dam site was thus chosen, under pressure for the supply of irrigation water, at a position that would be good for a reservoir, but not because it was good as a dam site.

Geology and Site Investigation

The high overall permeability of the volcanic bedrock was well known and well demonstrated by the large volumes of water that could be pumped from it with little depression of the water table, which was about 54 m below the river level near the dam site. River losses into the bedrock downstream of the site were about 700 to 1400 1/s, while upstream, the losses were negligible due to the perched water table. Drill holes made during the investigation lost their drilling water and one NX size hole in the right abutment had 30,000 m^3 of water injected during 15 days without filling it. Water levels in other boreholes in the right abutment rose during the injection and dropped abruptly when it stopped. A borehole camera inspection showed numerous cracks and joints of apparently random orientation, with a widest measured crack of 43 mm and many others in the region of 2 to 12 mm width.

A trial grouting programme was carried out on the left abutment during 1969. There were large grout takes in several holes and the take in just two of them exceeded the estimated quantities for the entire programme. Further boreholes were put down in 1970 to check on the effectiveness of the grouting. After a further review of the site investigation, it was decided that the rock forming the sides of the gorge was too open for efficient grouting and that a cut-off should be formed over the first 21 m depth with a trench cut into the rhyolite to be backfilled with core material and with the grout curtain continued down from the bottom of the trench. This trench would not be needed over the lower height of 23 m above the streambed, but would be required for both abutments above this level, the trench being taken up to crest level on the left side then along for 150 m into the left side and 300 m along to, and past, the spillway on the right.

Dam Construction

Presumably because at other dam sites the bedrock usually provides an impervious material, the dam was designed to have a cut-off trench with 1 on 2 side slopes, taken down through the 30 m of alluvium that fills the valley bottom to give a 9 m width of exposed bedrock so as to form a water-tight contact with core material that was to be used to fill the trench. This 98-wide strip was to be carried along the entire length of the centre-line of the dam, up the valley sides and along the bottom of the special trenches cut into the bedrock forming the sides of the valley/gorge. To save the volume of excavation, the trenches were to be cut as narrow as possible, consistent with maintaining the 9 m-width across the bottom of the trench. In practice, the side slopes of these trenches were at ½ (horizontal) to 1 (vertical). Considering that it was widely known that the rhyolite was very permeable *en masse* with all its fissures and open joints, it is most surprising that this design was accepted. Presumably, it was being assumed that the grout curtain would form a complete cut-off below the core-filled trench under the main part of the dam, as well as in the narrow trenches in the abutments, and that the high hydraulic gradient over the narrow grout cap would not cause any trouble.

The normal method of construction was used across the valley floor, with a wide trench 30m deep through the foundation alluvium, excavated by machine with side slopes of about 1 on 2, to expose the bedrock over a width of 9 m. This width was needed to accommodate the drilling and grouting machines required to place the grout curtain cut-off. This was formed from three rows of holes, with the central row of holes, some more than 90m deep, placed 3m apart, using split spacing where the primary holes did not indicate a tight curtain. The outer rows of holes were spaced at 6m centres simply to contain grout from the central row which otherwise could flow freely through the many fissures in the rhyolite bedrock. The surface of this rhyolite was pressure cleaned, ready to receive the compacted silt core fill of the wide core that was placed over the grout cap, against the cleaned surface of the rock and in the wide trench against the alluvium without any filter, on the presumed assumption that the grading of the sandy gravelly alluvium would act as a natural filter for the silt core.

Environmental constraints prevented the opening of a quarry for rockfill, causing the silt core to be made very wide and supported by fairly small shoulders, as shown in Fig. 4.8. A selected zone of sand, gravel and cobbles was placed next to the core (2 in Fig. 4.8), almost like a filter, but not designed as such. In the downstream shoulder, there was then a zone of miscellaneous fill (3 in Fig. 4.8) with some supporting rockfill (5 in Fig. 4.8), obtained from upstream, in the reservoir area. The silt was won by arrangements with the farmers by extraction below their land. Agricultural top soil was carefully removed and stock piled so that layers of silt could be taken off by scrapers and transported to the dam, where a special chute was constructed to convey the won silt taken from the plain at the top of the gorge, down to the bottom of the cut-off trench for placement. After extraction, the fields were reconstructed from the stock piled top soil so that the farmers could get back into production with minimum delay. The silt was placed in 15cm layers and compacted by 12 passes of a 6 t/m sheep's foot roller, at water contents intended to be 0.5 to 1.5% below optimum, producing a very strong and brittle fill. In the 30m-deep cut-off trench, the silt layers were simply put and compacted against the sides of the wide excavation, in direct contact with the natural alluvium that consisted of sand, gravel and cobbles.

During the excavation of the 21m deep, narrow cut-off trench in the right abutment, wide fissures were encountered. Two very large fissures were exposed, about 25m apart crossing the trench and one, about 1.2m wide, was explored by a geologist who was able to get a distance of about 30m upstream, downstream and below the trench floor. The second large fissure was not quite wide enough for a man to get in. There were also several large fissures exposed in the right wall of the canyon some 200 to 400m upstream of the dam. Construction of the grout curtain below the floor of the right trench caused some visible leaks coming from some of the exposed fissures. While the men were caulking these leaks, they saw numerous bats flying out of the cracks.

Similar experiences do not seem to have been reported in relation to the left abutment and it is possible that the rhyolite of the right abutment was more open. A letter written by the contractor contained the passage, "the rock on the right abutment was more fractured than that on the left abutment and there were more fissures in the key trench in the right abutment than in the left abutment key trench".

At the abutments, the rock surface was carefully prepared to take the core material, which was wetted slightly, i.e. about 0.5% below optimum for a 0.6m-wide zone in contact with the rock. Preparation of the rock surface was carried out 1 to 3 m ahead of the advancing fill, following a practice that would be used with a sound, competent, impervious bedrock. Any soil remaining attached to the rock was removed by handwork and high pressure air jets, with all the accumulated material taken away by a rubber-tyred backhoe. As is normal practice, no metal roller or tracked vehicle was allowed to contact the rock surface for fear of damaging it, ignoring the fact it already contained numerous fissures. The 0.6m-wide wetter zone was compacted against the rock by the rubber-tyred plant and with hand-held pneumatic punners, as seen in Fig. 4.10. The contact surface might have been expected to have been prepared with a covering of mesh-reinforced concrete or at least a thick coating of gunite in order to present a smooth surface to the core. In practice, where there were clearly open fissures, they were covered with slush grout and the particularly wide ones were filled with concrete. In some cases, mounds of core material were constructed on the surface of the rising core to direct the concrete from ready mix wagons into these

Fig. 4.10 Hand compaction of 0.6m wide contact zone.

wide cracks. No provision had been made in the specification for this treatment so that the contractor could not charge for this extra work. The methods used had been worked out by the site staff who could see that placing the silt core fill over open cracks would lead to trouble. The embankment and surface grouting inspectors, without assistance from the project geologist, would designate which cracks were to be filled. Due to the stiffness of the grout mixes used and also because the concern of the field personnel was generally for the larger cracks, fissures less than 6 to 13mm wide were ignored. In general, the frequency of fissures increased as construction reached further up the abutments, so that the detailed work of filling the fissures took more and more time. This led to core construction catching up with the fissure treatment, and a stage was reached when the ends of the core had to be sloped down so as to give room for the crack fillers. Since the contractor was not being paid for this work, and the width of the numerous fissures decreased in the upper parts of the abutments, it was decided to discontinue the treatment at 46m below the top of the right abutment, and rely on the core fill being compacted directly against the fissured rock. No expressions of concern came from the Denver design office, so the site staff may well have felt that their efforts were not appreciated and it was high time to give up trying to fill all these small fissures. There may also have been problems with extra costs, but this aspect has not been publicly discussed.

As has been said, the specification was for a situation where there was competent rock. It provided stringent conditions for the zone of core material placed next to the cleaned rock surface. The specified water content was obtained by mixing wet and drier fill and it was compacted in 15cm lifts by rubber-tyred plant and in 7½ cm lifts by the hand-held air or petrol tampers, as mentioned above. If laboratory tests revealed that the specified moisture limits had been exceeded, the material was removed, reworked and replaced. If the tests showed that the specified density had not been achieved, the area was recompacted. In a length of cut-off trench of 490m, a total of 425 density and water content tests were made. The difficulty of the placing conditions was shown by the fact that 114 of these tests failed and required reworking the material in order to bring it within specification. Difficulty must also have been experienced due to variations of the optimum water content, which varied from 26% to 24%. The mean for the material being placed in the cut-off trench was 19.1% and the aim was to place it at 0.6% dry of optimum. The fact that so many areas were reworked shows the conscientious approach to this important task, even though the significance of the rock fissures was being ignored.

A century ago, it was British practice to form below-ground cut-offs by trenches filled with core material, usually puddle clay. The trenches were excavated with vertical sides, supported by timbering. The problem of fissures in the bedrock was well understood, and provision made in the specification for concrete walls to be cast against the cleaned rock surface. The following extract is from the specification for the Winterbourne Reservoir dated 1885:

> If it should be considered necessary by the Engineer, the Contractor shall protect the sides of the puddle trench, or portions of the same, by means of concrete walls on either side of the puddle, and the contractor shall be paid for the work at the price per cubic yard given for the same in the Schedule of

> Prices attached to his Tender. The concrete walls shall not be less than 18 inches (0.46m) thick, built plumb and parallel with the centre line of the trench, smoothly faced against the puddle, and shall quite fill up all fissures in the rock and all irregularities in the sides of the trench.

Constructing separate concrete walls on either side of the narrow puddle clay fill was not easy and it became the practice to backfill the cut-off trench completely with concrete. Such a procedure at Teton could have provided a non-erodible, reasonably watertight cut-off above the grout curtain. Had the concrete work been extended outwards to give a thick, reinforced covering to all the rock over the contact surface with the core material, the potential for erosion of the silty core could have been substantially reduced. The extra cost for construction, however, might have prevented the reservoir from being economically viable.

Analysis of the Failure

The positions of the whirlpool in the reservoir and the swallow hole in the downstream slope were both in line with the position in the right key trench where the work of filling the rock fissures with grout or concrete had been discontinued. Water from the filling reservoir could pass through the rock fissures so that full reservoir pressure could come to bear on the upstream side of the silt fill in the narrow cut-off trench and the upstream side of the narrow grout cap. The hydraulic gradient across this was great and it could be expected that seepage would occur over the cap, beginning to erode the silt and carry it into the fissures on the downstream side. The ready connection between the reservoir water and the rock downstream of the dam centre-line was shown by the rise of water level in boreholes in the downstream part of the rock in the valley side as the reservoir level was raised by the impounding.

An analysis of the shape of the narrow cut-off trench showed that vertical total pressure in the silt would be less than expected from the height of fill above it, near the bottom of the trench, due to load being transferred to the sides of the trench, so that it could be exceeded by the pressure of the water from the reservoir causing the development of hydraulic fracture. Detailed finite element analyses clearly showed that an arching action in the narrow key trench could cause the vertical stresses to be less than the pressure of the reservoir water at the position where failure initiated—with the height of water in the reservoir at that time. As a check measure, hydraulic fracture tests were made in the remaining portions of the core from boreholes. This is not a good method because the formation of a borehole can completely upset the stress conditions it is intended to measure. It would have been much better to carry out hydraulic fracture tests from existing hydraulic two tube piezometers, had any been installed. The USBR developed the twin tube hydraulic piezometer and fitted them to a large number of their dams. Walker and Dhean (1948) describe 10 years of pore pressure measurements made by the Bureau in their dams, yet when it came to Teton, they appear to have decided that they had sufficient information about the development of pore pressures in embankment dams and that money could be saved by *not* installing any instrumentation in this dam. The result was that there is no record of the developments of pore pressures within the dam fill during construction nor during reservoir impounding prior to failure.

It was argued that the shapes of the left abutment and key trench were extremely similar to those of the failed right side, so that pre-failure conditions on the right could be expected to be similar to those existing on the still intact left side. Accordingly, boreholes were put down through the core on the left side and into the narrow key trench. A major difficulty is that the pressure of the water used in making the borehole is likely to cause hydraulic fracture. Not only has the fill to withstand the pressure of the vertical head of water in the hole, but the circulating pump can develop high pressures when any obstruction occurs in the return flow up the borehole. Replacement of drill water by air can make matters worse because of the high pumping rate necessary to blow the cuttings out of the hole. Any blockage can cause the development of really high pressures. It is small wonder that hydraulic fracture commonly occurs as the holes are drilled. Sherard (1973) has given several examples where flush water was suddenly lost when a certain depth below crest had been reached, leading to the concept of a horizontal crack through the core at that particular depth.

The holes drilled in the left core fill at Teton were no exception. One hole lost more than 13 m^3 of water at a depth of 30 m. Other holes drilled with air were fractured by the air pressure, as was revealed when the large excavation was made. The positions of the holes that were drilled for hydraulic fracture tests are shown by Fig. 4.11 and it was concluded that the few successful hydraulic fracture tests made in these holes show fracture pressures slightly higher that the pressures produced at those levels by the reservoir water. Subsequent excavation showed that the formed cracks were mainly vertical and transverse across the cut-off trench. This suggests that in the trench, the minor principal stress direction was along the trench. A comparison of the hydraulic fracture pressures measured within the cut-off trench (hole HF-5) with those measured in the core (HF-15) has shown that the minimum principal stress in the trench was about 30% less than that at corresponding depth in the main core: an indication of weight transfer from the core to the

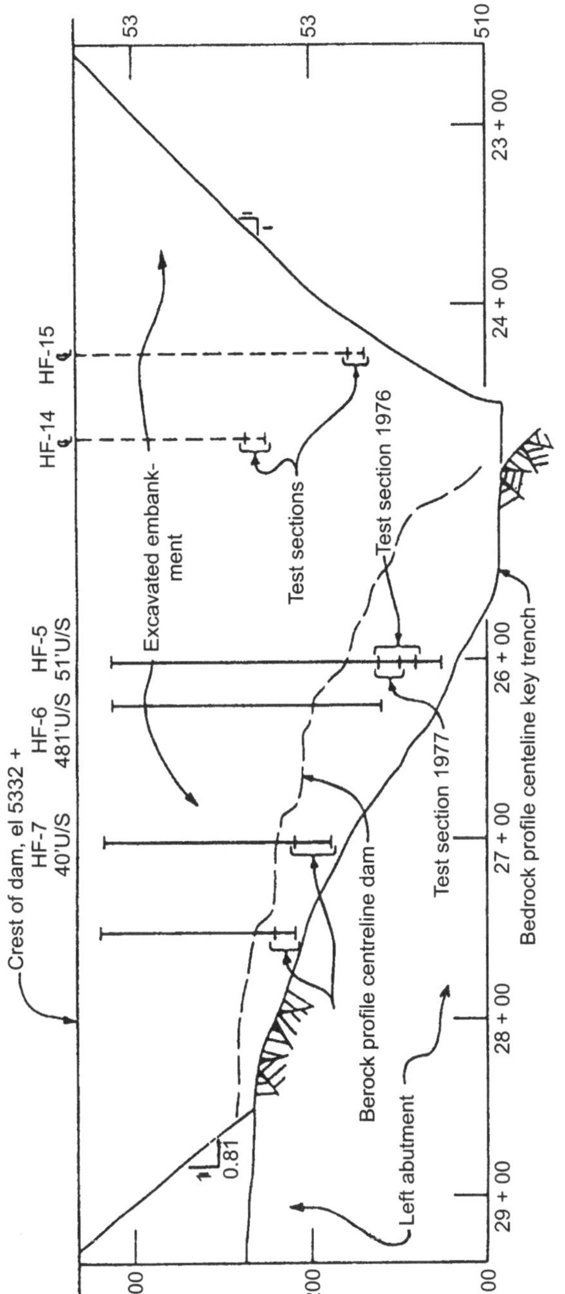

Fig. 4.11 Position of holes drilled for hydraulic fracture tests.

trench walls. The data from hole HF-5 was considered to be indicative of the fact that there were zones in the embankment where the total stress was low enough to allow hydraulic fracturing within the cut-off trenches at the time of the dam failure. At that time, the reservoir level was at 8.8 m above the indicated initial fracture pressure for HF-5 and 10 to 11 m above the pressure at which the fracture closed.

Some Consideration of the Properties of the Silt Core

Mineralogy tests have shown that the Aeolian silt contains particles that probably originated as volcanic ash. The volume of the core fill was given as 3,965,466 m^3, more than half the volume of the whole dam. Average placement water content was 18.6%: 1.3% dry of optimum, with an average dry density of 15.6 kN/m^3. The average specific gravity of the silt was 2.62, so that the average void ratio would be 0.646, giving a degree of saturation of 75%. The average bulk density of 18.5 kN/m^3 was not particularly high considering that compaction, seen in Fig. 4.12, was by 12 passes of a 6 tonne/m sheep's foot roller on every 15 cm layer. This heavy compaction of the silt, dry of optimum, produced a strong fill that must have had very low pore pressures. This fact is well illustrated by the vertical face of the core on the left side of the breach, as can be seen clearly in Fig. 4.13.

The recommendations made by Proctor (1933) that the fill should be placed at an optimum water content was to overcome the problem of collapse settlement on the wetting of fill that had been placed too dry: His compaction energy of 596 kJ/m^3 is lower than that provided by modern placing machines, and has been increased to 1,136 kJ/m^3 in the USBR compaction test E-11 (Bureau of Reclamation 1974). The optimum water content found from this test could have been expected to provide an ideal workability, but the specification called for a placement within the range 3.5% dry to 1% wet of this optimum. The inspectors found it extremely difficult to judge the water content of the silt coming from the borrow area and requested numerous water content tests that were made in the project laboratory using a microwave oven. Although the silt was of uniform appearance, it varied greatly in both natural and optimum water contents. Results from approximately 2,167 in situ density tests exhibited an average compaction of 98.6% of optimum dry density and an average placement water content 1.2% dry of optimum. The saturated silt complies with the principle of effective stress and Penman (1953) has shown that when undrained, it behaves as a $\phi = 0$ material in terms of total stresses. Shear strains can cause a reduction of pore pressure of about 600 kN/m^2 in a dense sample: if confining pressures were too low, this could lead to cavitation and drained behaviour. Unsaturated silt does not comply with the principle of effective stress. It exhibits high pore

Fig. 4.12 Compaction by sheep's foot roller.

Fig. 4.13 Vertical face of silt core on the left side of the breach.

suctions, producing a large effective stress that gives the material rigidity. A reduction of effective stress caused by wetting while under a confining pressure results in compression instead of expansion.

Kane (1973) has studied the behaviour of natural samples of loess with water contents ranging from air-dry to saturated, under confined compression. The use of a pressure plate apparatus enabled him to measure pore suctions of 2,000 kN/m^2 in the drier samples. Kane found that the drier material was very rigid and showed negligible compression under increasing vertical total stress upto 700 kN/m^2. He attributed the rigidity to bond strength between the silt particles provided by a thin coating of clay. Examination under very high magnification of thin sections cut from samples of the Teton core has shown that there are thin, nearly continuous 1 to 5 μm thick oriented clay coatings on most skeleton grains. Maswoswe (1985) has extended this work, using a compacted sandy silt under confined compression in an oedometer. Top and bottom porous plates were kept dry and pore suctions were not measured. Compaction, at a water content slightly dry of Proctor optimum, was adjusted to produce various void ratios. At an initial voids ratio of 0.65, the material showed little compression under a vertical stress upto 430 kN/m^2; however, wetting the sample under a variety of vertical stresses caused settlements to the compression curve of a wet specimen, as shown by Fig. 4.14. In another test, after the total vertical stress had been

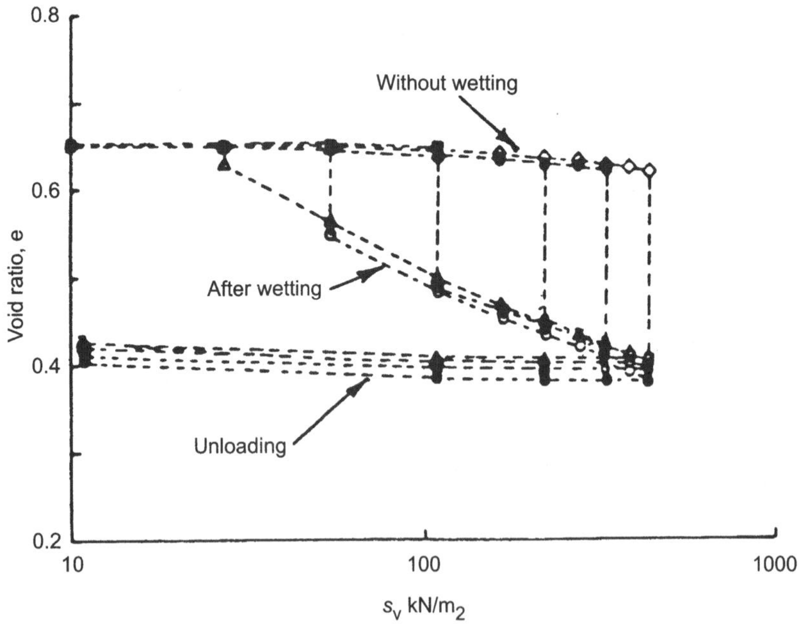

Fig. 4.14 Collapse settlement caused by wetting.

increased to 430 kN/m^2, the sample was held at constant volume as it was wetted and the vertical stress fell to 26 kN/m^2, as is shown in Fig. 4.15.

Leonards and Davidson (1984) found from a detailed study of the compaction control tests made during the construction of Teton, that there were layers of silt fill placed between 49 and 53 m above the streambed, dry of optimum by up to 2.2%, at the horizontal positions where failure initiated. Oedometer tests on laboratory compacted samples of the core silt were loaded to a vertical total stress of 550 kN/m^2 before being soaked with water. The ensuing collapse settlement depended on the placement water content; at 2.2% below optimum the volume reduction on wetting was nearly 2%. They also point out that this volume reduction would reduce the total vertical stress, permitting hydraulic fracture under reservoir head. All the test results described above support this argument. The layers described by Leonards and Davidson—placed dry and extending into the narrow key trench—could be subjected to the reservoir water at their upstream edges. Being only a fraction of the width of the main core, and with the rigid core fill above them supported by the trench walls, collapse settlement would have a more pronounced effect in reducing total stresses. This could lead to the more rapid ingress of reservoir water and the collapse of the entire width of the layer right through the narrow cut-off trench. Such an event, as suggested by Leonards and Davidson, could permit sufficient flow from upstream to downstream of the cut-off trench so as to remove the silt into the fissures and joints of the downstream bedrock.

The Large Excavation

Following the idea that the shape of the left abutment was similar to that of the failed right, the decision was taken to excavate the dam fill on the left side so that the conditions of the contact between silt core and bedrock could be examined and, in particular, the conditions in the key trench could be studied. This was a very important move and revealed a condition that was perhaps the most important finding resulting from the failure.

A very large volume of fill had to be removed to maintain a stable slope on the end of the dam. The excavation was of comparative size to the stabilised breach on the right side, leaving only a pyramid-shaped central piece of the dam still standing on the valley floor, as seen in Fig. 4.16. The excavation was carried out very carefully, making in situ tests at all stages in an attempt to find any irregularities. As the streambed level was approached, the time and money allocated for this exercise had expired, and preparations were begun for finishing the excavation work. A small group interested in the research aspect begged for just another

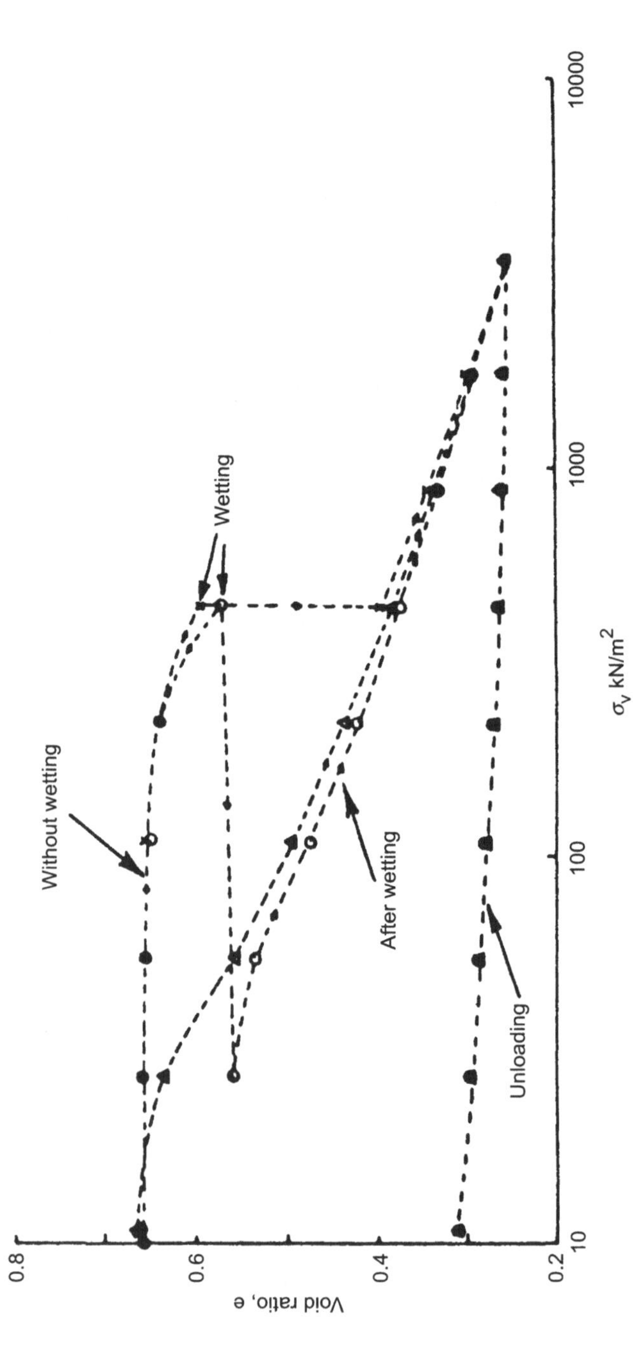

Fig. 4.15 Consolidation after wetting and collapse under constant volume reducing the vertical stress from 430 to 26 kN/m^2.

Fig. 4.16 Pyramid-shaped remainder of dam.

day of excavation. In the morning, when they returned for their last day of digging, water was found seeping from the core silt in a horizontal line near the bottom of the deep excavation. How could there be so much water coming out of the silt, and that too in a more or less level line?

This finding resulted in authorisation of much more work. To begin with, an adit was excavated parallel to the dam axis about 33m long, just above the wet seam so that short boreholes could be made into the seam to study its properties and try to find its extent. This wet seam extended beyond the end of the adit and it was later detected across the whole length of the core by boreholes put down from the surface. Figure 4.17 shows the extent of the wet seams. Samples showed that it was, of course, much wetter than the adjoining fill; it had a lower density and was, therefore, more permeable. It was just a little above the shut-down surface left during the winter of 1974–5. Had it been below that surface, the argument could be made that it had been caused by frost heave. A great deal of work was done to check all the engineers' field notebooks in order to ensure that the wet seams were, in fat, above the winter shut-down surface, even to the extent of X-raying the books in a search for any alteration that might have been made. Construction photographs

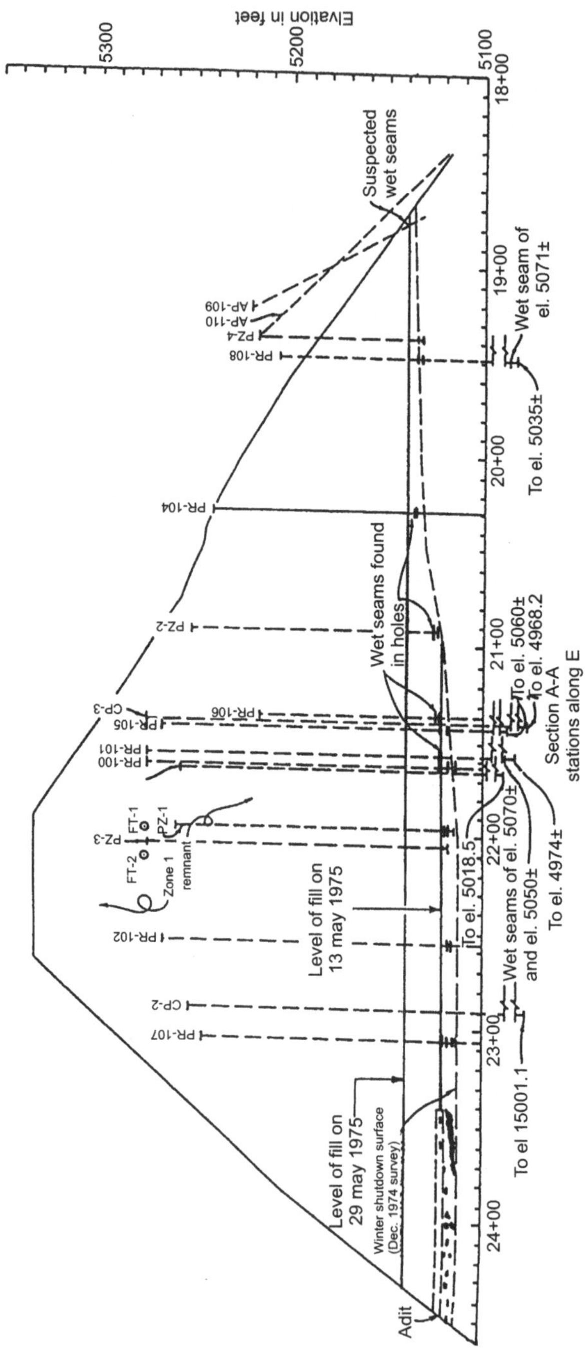

Fig. 4.17 Extent of wet seams

confirmed that it had been a wet spring so that when construction started again in 1975, there were pictures of pools of water on the surface of the fill and suggestions were made that excess water could have been built into the fill at that level. But that level was 61m below the dam crest.

Causes of the Wet Seams

Strong views were held that the wet seams were the result of ice lenses, even after it had been shown convincingly that they were all above the 1974–5 shut-down level, where the surface of the fill had been exposed to the severe frosts of winter that could well have caused the formation of ice lenses. Ice lenses develop because the vapour pressure of ice is lower than that of water, so when pore water is frozen, the lower pressure attracts liquid water from below. The silt fill had just the right pore sizes to maintain the lower pore pressure yet was sufficiently permeable to enable pore water from below to rise to the frozen level and add to the thickness of the forming ice. During the months of winter, there was opportunity for reasonable thicknesses of ice lenses to have formed and when they melted, they could have released water to form wet seams. But below the 1974-5 surface, no signs of wet seams were found. If ice lenses had formed, they would have been fairly close to the surface and any wet material would have been removed in the spring when all wetter or damaged fill was scraped off in preparations for continuing placement.

Another approach was that perhaps frozen fill from the borrow pits had been placed to form the first layers of fill above the 1974-5 level and also the wet spring could have caused these layers to be particularly wet. So, had the contractor been at fault in placing wet fill that had not been detected by the inspectors? Records showed that during the early weeks of the new 1975 placing season, there had been a shortage of inspectors, so possibly some wet fill had slipped through. Another suggestion was that particularly dry fill had been placed during the beginning of the season. Permission had been given to use fill at water contents as low as 3.7% below optimum and such fill could have been further dried by wind and sun during placement in thin layers. Tests made by the USBR in their own laboratories as far back as 1942 had shown that fill placed drier than about 2% below optimum suffered collapse settlement when subsequently wetted. The fill at Teton placed at this low water content could have developed layers that collapsed as a result of heavy rain percolating down into the new fill which resulted into the formation of wet seams. Tests on the wet seams showed that they had dry densities as low as 85% of optimum. This would make them more permeable and after reservoir impoundment began, water from the reservoir could have entered from the upstream side.

Most Likely Cause of Failure

The late Dr James Sherard (Sherard 1987) gave the most feasible explanation for the wet seams. He pointed out that the portion of the dam containing them had been constructed during the Spring of 1975, and the wet seams were first discovered in October 1977, about 30 months later. The dam had reached full height in November 1975, so the wet seams had been under the pressure of 61m height of fill for about 23 months. During that time, any excess water added to the fill by contractual error or rainfall would have been pushed out by consolidation and sucked up by capillarity into the adjacent embankment layers. He argued that the water must have come from the reservoir in 1976 after the water level had risen above the wet seam levels.

The reservoir had risen to the level of the seams by December 1975 and continued to rise a further 56m by the day of the failure on 5 June 1976, so that it had been above the seams levels for more than 5 months. General seepage into the core could not account for discrete seams of wet material: something else must have occurred to permit water to enter along a seam. That something else—Sherard proposed—was hydraulic fracture.

Sherard further suggested that the dam above the streambed level could have developed an arch structure over the wide cut-off trench that had been taken down below the riverbed to the rocks and backfilled with the silt core fill. In discussion with Penman, Sherard agreed that after the trench had been filled to riverbed level in the summer of 1974 and the pumps had been turned off, water would begin to go into the fill, a process greatly accelerated when impounding began. This phenomenon could cause the collapse of the settlements, thereby reducing the vertical total stress under the developing arch. Sherard's sketches illustrating this mechanism are reproduced in Fig. 4.18. Position 'c', at riverbed level could move down due to collapse compression of the fill below, while positions 'a', 'b', 'e' and 'f', being on original ground, would not suffer any appreciable settlement. Position 'd', just under the developing arch, would not settle much, reducing the compressive stress under 'd' until, as the reservoir level rose, a stage could be reached when the water could enter the core fill in a thin concentrated leak by hydraulic fracture. The horizontal crack would, in effect, be jacked open by the increasing water pressure, but because the silt lacks cohesion, the roof of the crack would collapse in the form of loose silt falling through water, so filling the crack with loose silt. As the crack propagates through the whole width of the core, it becomes filled with sufficient loose silt as to reduce its permeability, so that when it reached the downstream edge of the core, resistance to flow was sufficient to prevent rapid, eroding flow. Sherard

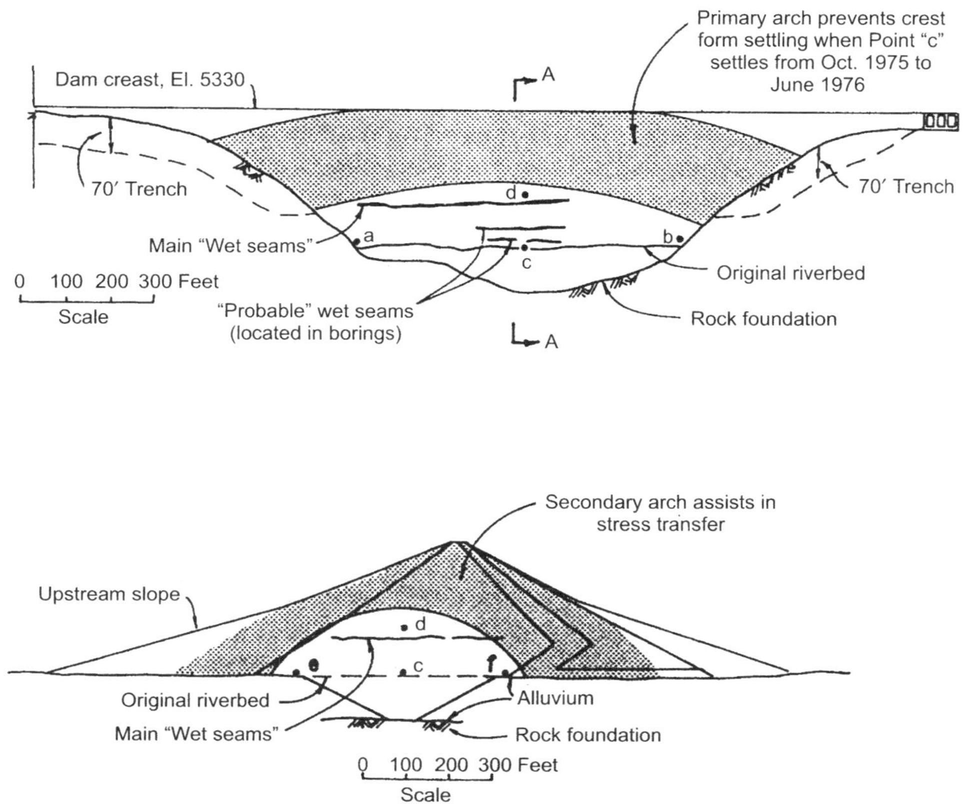

Fig. 4.18 Sherard's drawing illustrating the mechanism of hydraulic fracture.

(1985) had witnessed such horizontal cracks filled with loose silt in the core of the Manicouagan 3 Dam in Quebec.

Evidence of the strength of the compacted silt core was shown by the near vertical face left when the breach occurred, as seen in Fig. 4.13. This near vertical 90cm-high cliff was stable for months, forming the left side of the breach, and remained so until it was brought down by machines during the safety stabilizing work later on. The fact that an arch can form in this situation is shown by the photographs in Figs. 4.4 and 4.5. It can be said that the most interesting aspect and lesson learnt from the Teton failure was the discovery and study made of the wet seams.

Other Causes of the Failure

Professor Victor de Mello said that there were a dozen ways in which the Teton dam could have failed, and it chose only a few! The failure taught no new technical lessons: the design and construction methods used for embankment dams were quite adequate for the construction of a safe dam, but many aspects had been ignored. One major factor was bureaucratic. The dam design section of USBR was in their headquarters in Denver, Colorado, and when any of the design engineers went to the site of a dam, it involved a claim for travelling and hotel expenses. The administrative people, chained to their desks in Denver, were envious of the engineers going off into the country and living it up in fine hotels. So they introduced a rule that design engineers had to design in their Denver offices and leave the site staff to build the dam as they had designed it. Although the site for Teton dam was not too far from Denver, the design team only got the opportunity of visiting the site twice; both during the early planning stages, and never during construction of the dam. The overall pervious nature of the bedrock was well known from the high yields of the deep wells used by the farmers to irrigate their crops; a situation reinforced by the experience of the trial grouting programme, when grout and water had been pumped into trial boreholes without obtaining a seal or in filling the hole with the vast amount of injected water.

Despite this knowledge, the designers went ahead with a design suitable for a tight valley with impervious bedrock. The main wide cut-off trench was taken down below the river level to reach the fractured bedrock, which was exposed over a 9m width to accept the base of the silt core that was to be compacted directly on to the rock surface! The grouting trials had shown that it would be impossible to form a seal in the abutments, so the design called for a 21m deep cut-off trench to be incised into both the abutments. To reduce costs, the trench was to be made as narrow as possible, consistent with forming a 9m width at the base to take the drilling and grouting equipment that was to form the grout curtain under the entire length of the dam. It was to be a three-line curtain; the two outer rows were to try to retain the grout placed from the central row of holes, but not to form an impervious curtain themselves, so it could not be considered as a three-row grout curtain, but actually, only a single row curtain. Hence, the grout cap was only over the central row, and was relatively narrow. While grouting was being carried out in one area, bats were driven from the wide fissures in the bedrock wide enough for them to enter and explore for considerable distances under the zone for the grout curtain. None of this information appears to have got back to the design team in Denver, who were probably busy with another dam by then.

The design for the Teton dam called for the silt in the narrow cut-off trenches to be compacted directly against the fissured sides, but the site staff, realizing the futility of this venture, by their own initiative, grouted up the wider fissures as the level of fill reached them. Many fissures were so wide that concrete had to be used rather than just grout. They used to make mounds of the silt fill so that ready mix lorries could back up along the surface of the core and discharge their concrete behind the mounds so that it ran straight into the fissure. When several lorry loads were needed, accelerators would be added so that the mix took an initial set to seal the fissure, enabling the next load to fill it up. As the level of the fill rose, the fissures became smaller and more frequent, requiring more time to seal them. This caused the fill level to begin overtaking the sealing work, till the stage was reached when the sealing of each small crack had to be abandoned. It is not clear what allowance had been made for payment of all this extra work, but it is possible that there was no extra payment and that the site staff were doing it because they felt it to be necessary. In general, the cracks higher up were only a few mm wide, so it might have been felt that they would be sealed by the silt compacted against them. However, failure initiated at the level and place where this sealing of the cracks had been abandoned. It would appear that the reservoir water had no trouble in reaching the grout cap in the bottom of the narrow side trench, and clearly once it crossed the cap, it had no difficulty in escaping into the multitude of fissures on the downstream side that led back into the valley itself. The passage over the grout cap would have been assisted by the arching action of the compacted silt in the narrow trench, causing a reduction of vertical stress of the silt on to the grout cap. As part of the post-failure study, a finite element analysis indicated that the normal stresses in the silt in the trench, as shown in Fig. 4.19, became less than the reservoir water pressure as the level of the reservoir approached the level at which failure occurred. This effect was accentuated by the fact that the silt compacted dry of optimum, would suffer collapse settlement on wetting, thus, still further reducing the vertical pressure of silt on the

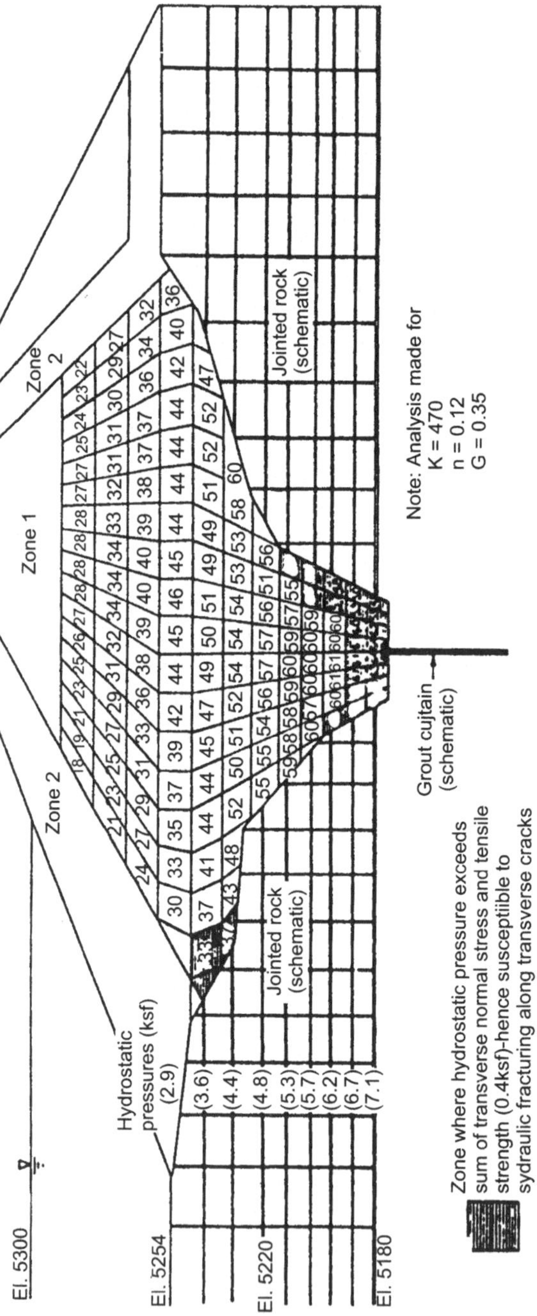

Fig. 4.19 Computed values of normal stress on a transverse section where failure initiated.

grout cap, permitting of hydraulic fracture. Once water began to escape over the narrow grout cap, the silt would rapidly erode, quickly leading to the complete failure that was observed.

PANSHET AND KHADAKWASLA DAMS

Panshet dam was constructed in the years 1957–61 as a rolled earthfill dam across River Mutha in Maharashtra, India. This dam was located 22 km upstream of an old existing masonry dam Khadakwasla, which in turn, was about 20 km upstream of the city of Pune, which is spread on either banks of the river Mutha. The rivers Mula and Mutha join together just outside the city of Pune. The river is called Mula-Mutha after the confluence as it passes through the middle of the city.

Both Panshet and Khadakwasla dams failed in July 1961 after an unprecedented and heavy rainfall, which started on 25 June and was recorded to be 32 cm on 26 June 1961. The rainfall continued intermittently till 1 July 1961 and then went on incessantly, reaching maximum intensity on 9 July, which recorded a volume of 21cm. On 10 July, there was heavy downpour of the order of 42cm when trouble started at Panshet dam located about 35km from the city of Pune, and the dam breached on 12 July. The breach of Khadakwasla dam, located about 16 km upstream of the city of Pune, took place about seven hours after the breach of Panshet dam. Both these breaches took place in the daytime and caused damage to 4,420 houses in the urban area and 630 houses in the rural area, affecting 95,000 persons.

Causes of Failure of Panshet and Khadakwasla Dams

A number of opinions were put forward to explain the reason(s) of failures of both these dams. These were either controversial or complementary.

The examination of the remaining portions of the dams to find out the causes of failure was a challenging task for the engineers. These experts differed in views and putting forth the reasons. The concerned state government, the Government of Maharashtra, decided to order an inquiry and appointed a one-man Commission of Inquiry (1961) to inquire into the failure and report on:

(a) The causes of the failure of the two dams and the circumstances in which the failure occurred; and
(b) The adequacy of the action taken before, during and after the disaster to mitigate the consequences thereof.

This chapter is mostly based on the report and discusses the causes of failure of the dams and the reasons for the same.

PANSHET DAM

The Panshet dam was a zoned rolled earth dam of 56m height with elevation of top of the dam as EL 639.17m, full reservoir level and maximum water level EL 636.73m and normal storage level upto EL 628.50. The volume of earth dam upto an elevation of 636.13m was worked as 3.51 million cubic meters. The design included an outlet conduit capable of discharging 28.32 cubic meters per second located in a 21.34m-deep trench excavated in hard rock under the left bank of the earth dam. The designed cross section of the dam is shown in Fig. 4.20 while Figs. 4.21 and 4.22 give the plan and longitudinal section of Panshet dam. The construction of the dam was planned in five working seasons from 1957–58 to 1961–62. This schedule did not show the construction plan of the outlet conduit and control works. In the first two working seasons, the good progress made led to a revision of the programme and completion of dam in four seasons, that is before 1961 monsoon. The delay at various stages and poor management by the construction agency delayed the construction of conduit. The work of backfilling the trench over the completed conduit proceeded slowly. The material to be filled up in the gaps, between the conduit and excavated rock, and upto a meter above the arch of the conduit, was to be placed and compacted manually. On 19 June 1961, earthwork had to be discontinued due to an early onset of monsoon.

On the eve of the failure, the following important items of work remained incomplete:

1. The height of the dam in the temporary waste weir section was constructed to EL 632.46 m as against the designed height of R.L. 633.98.
2. The level of impervious zone in this section was EL 628.80m as against the designed level of EL 632.76m.
3. The downstream casing in this section had reached the level EL 629.10m against the designed level of EL 633.98m.
4. Emergency gates were not erected.
5. The hoisting arrangements were incomplete, as a result of which the gates could not be operated. The service gates were kept suspended by wire ropes.
6. The floor of the conduit was not concreted.
7. The upstream pitching remained incomplete.
8. The downstream transition box and retaining wall were not constructed.

From 12 May to 19 June 1961, for about five weeks, progress of earthwork was about 1.22m per day, before the work was suspended due to heavy rains.

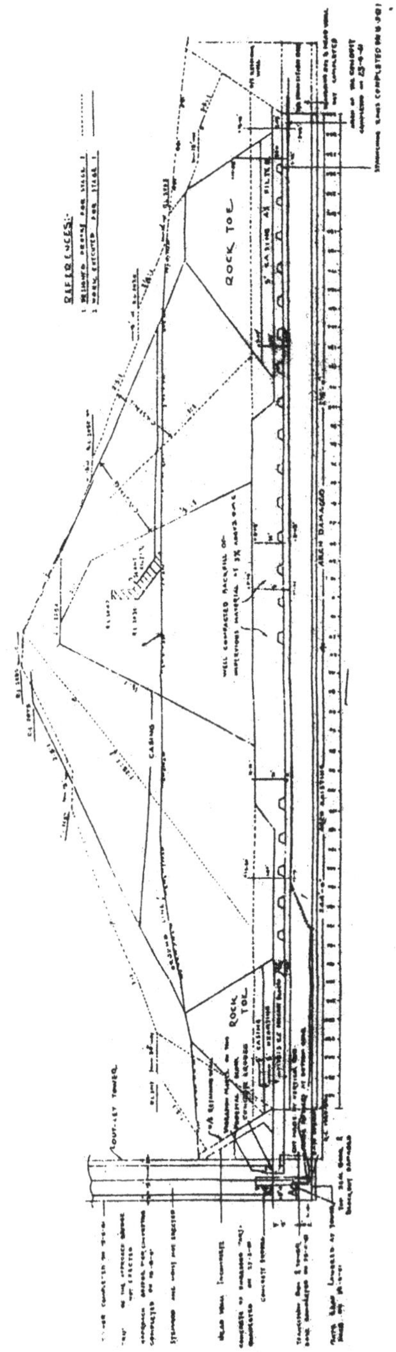

Fig. 4.20 Showing detail of incomplete dam section control tower and conduit after failure

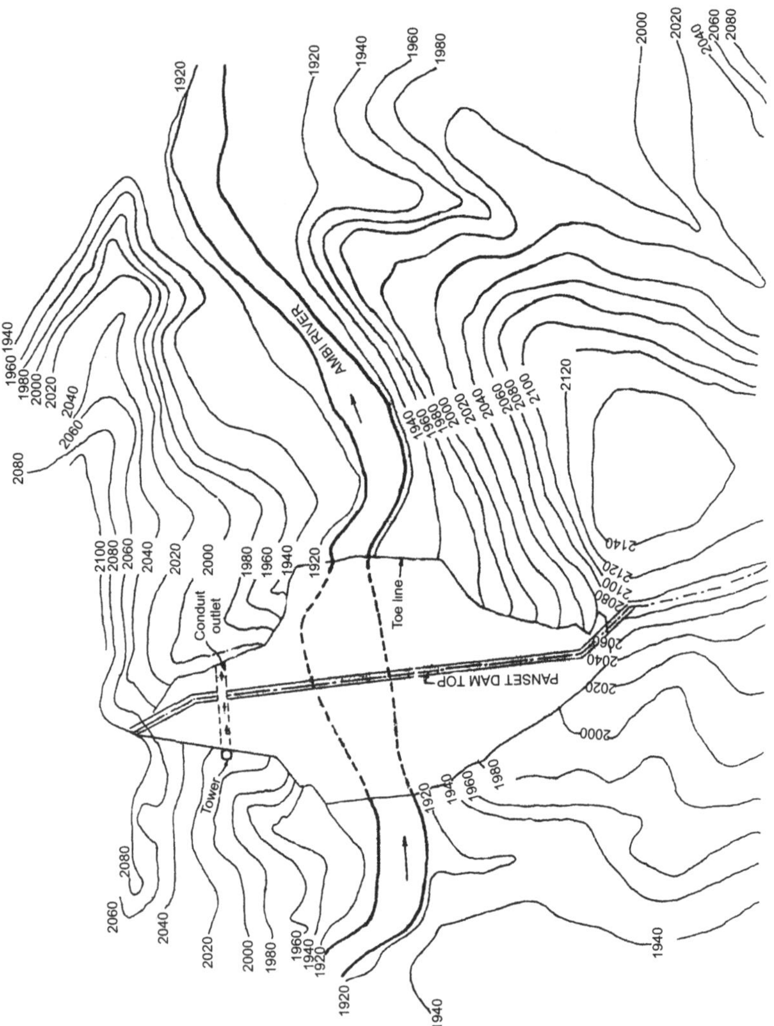

Fig. 4.21 Layout plan of Panshet dam (in feet)

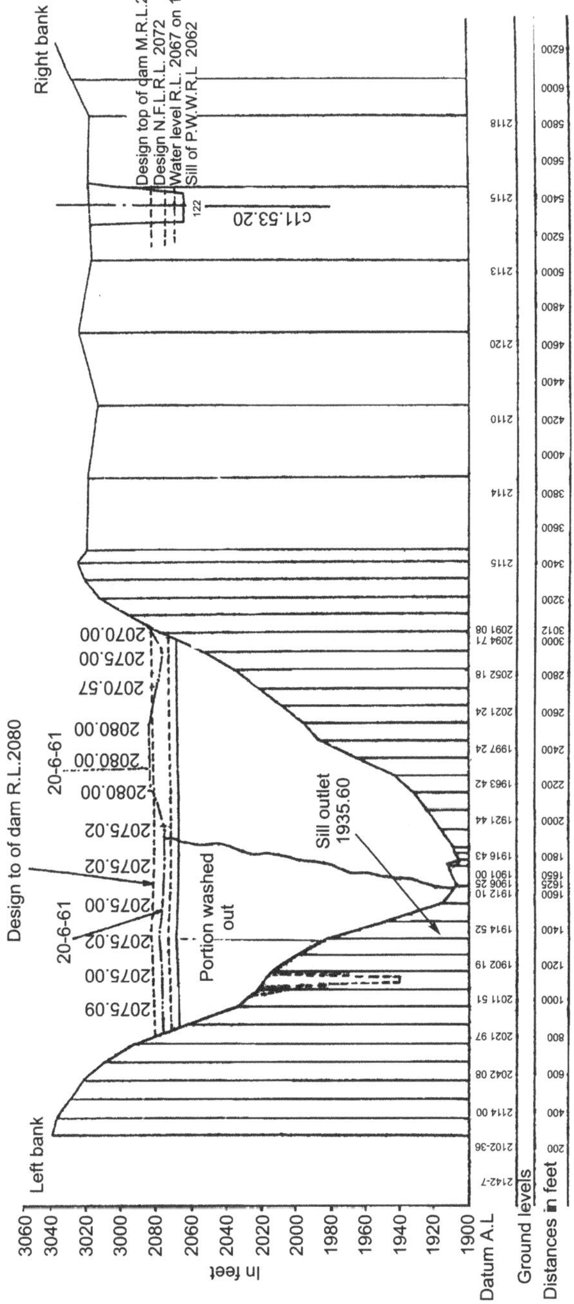

Fig. 4.22 Longitudinal section of Panshet dam showing dam, P.W.W. and T.W.W.

During the rains, there was a rise of 12m of water level in two days and after a gap of four days the level further rose by 18.3m. The storage further increased by 2.5m in the last four days, that is before the dam breached. The rapid construction followed by rapid filling of the reservoir is considered as one of the reasons of the failure of the dam. The graphical presentation as given in Fig. 4.23 shows the happenings of the last two days of the life of the dam, and Figs. 4.24 and 4.25 show the cracks that developed.

On July 10 1961, the lake was at EL 628.50 and heavy rains along with high winds of the order of 53 to 62 km per hour, created wave wash conditions. This damaged the portions of the dam on both the sides of the conduit gap, which had been completed just before the severe weather conditions and were left without rip rap on the upstream slopes and were protected only by galvanized sheets and jute bags filled with sand and freely draining material. An unsteady flow was also noticed at the downstream toe of the conduit on the same day, producing a hissing and loud thud noises and the water level was rising and falling 15 cm above and below the conduit top. At about 19.00 hrs, water was emerging from the downstream rock toe an discharge of the order of 0.14 to 0.28 cu.m per sec was estimated from the top of the dam in that rainy conditions, taking a position at 45.72m above and 91.44m away from the downstream leakage.

On July 11 1961, at 02.30 hr, settlement was observed on the upstream slope above the conduit. Transverse cracks, 9.14m apart along the axis of dam and 6.10m perpendicular to the axis, were noticed along the sides of the settled slope. At about 05.30 hr when the subsidence increased to 1.37m, the free board was reduced to 61 cm and then on to 91 cm. Attempts were made to raise the top of the embankment by placing jute bags filled with sand and gravelly materials. On July 11, the reservoir water level increased by 1.68m by 11.00 hr. At 17.30 hr, two more transverse cracks appeared about 30.5m apart at the top of the dam and on the downstream slope of the casing. The Indian Army, sappers and miners continued their struggle to keep the top of the dam above the lake level but the subsidence continued and finally the dam was overtopped at 06.30 hrs on 12 July 1961. The left portion of the dam disappeared upto the old river course and about 84m of the length of the conduit, on the downstream, disappeared, leaving 74.5m length in place even after the breach.

Causes of Failure

The observations of the two days before the failure of the Panshet dam and the manner in which it failed did not provide any straightforward

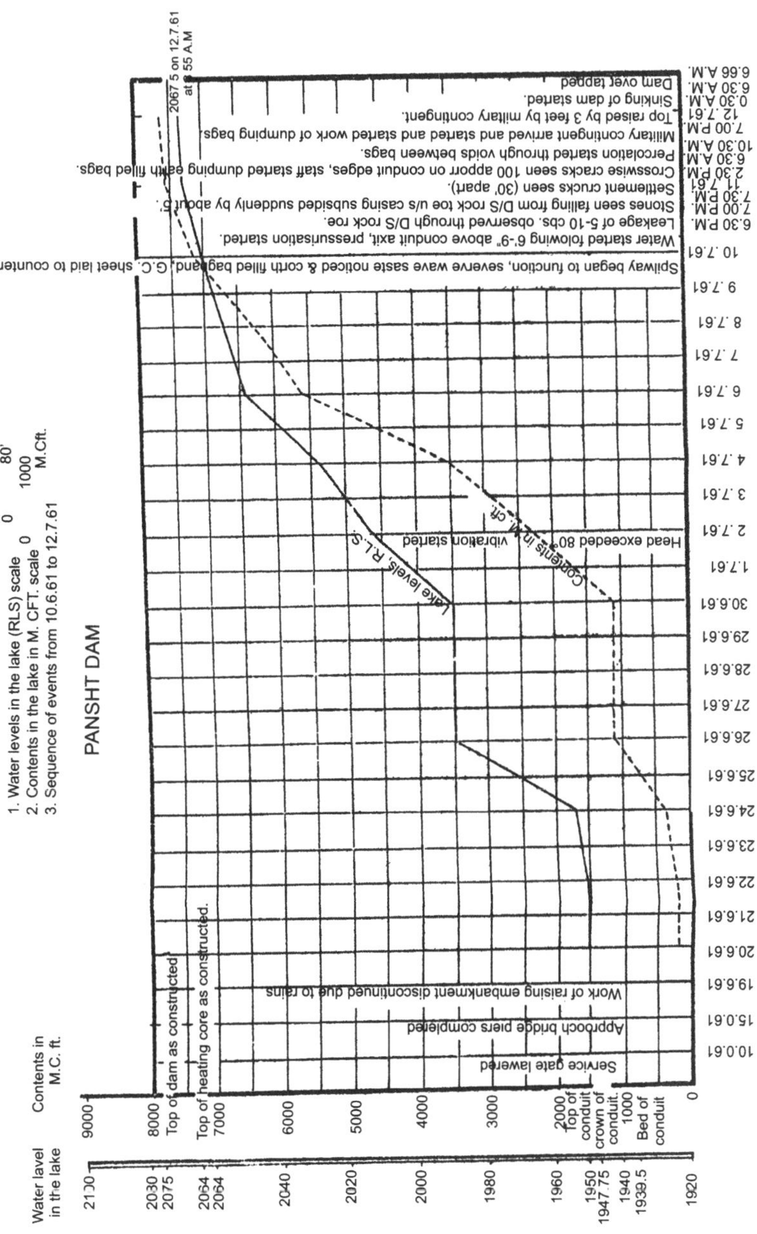

Fig. 4.23 Graphs for the period from 10 June 1961 to 12 July 1961

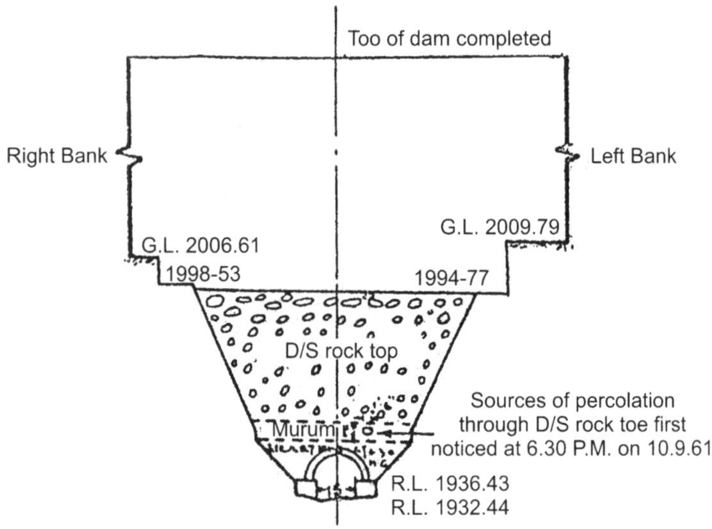

D/S El. evation
Showing location of per colation

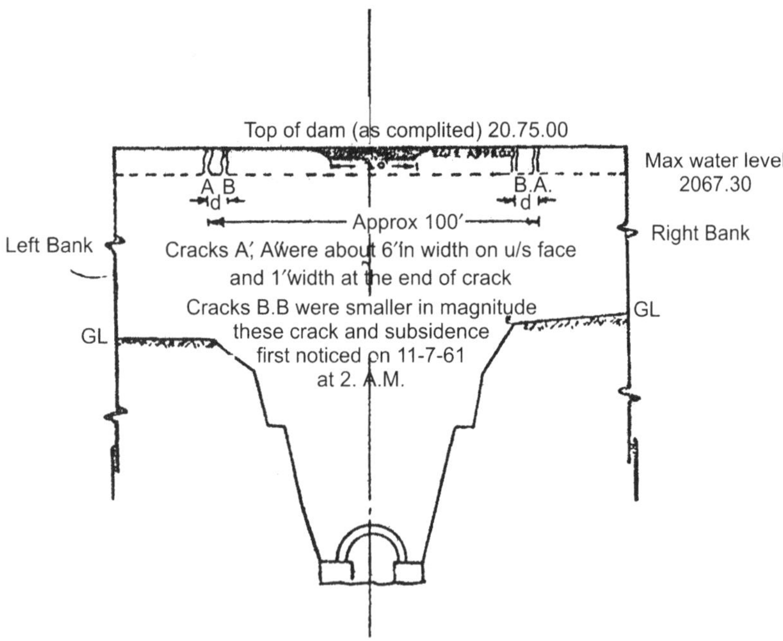

U/S elevation
showing subsidence & cracks in U/S casing.

Fig. 4.24 Panshet dam

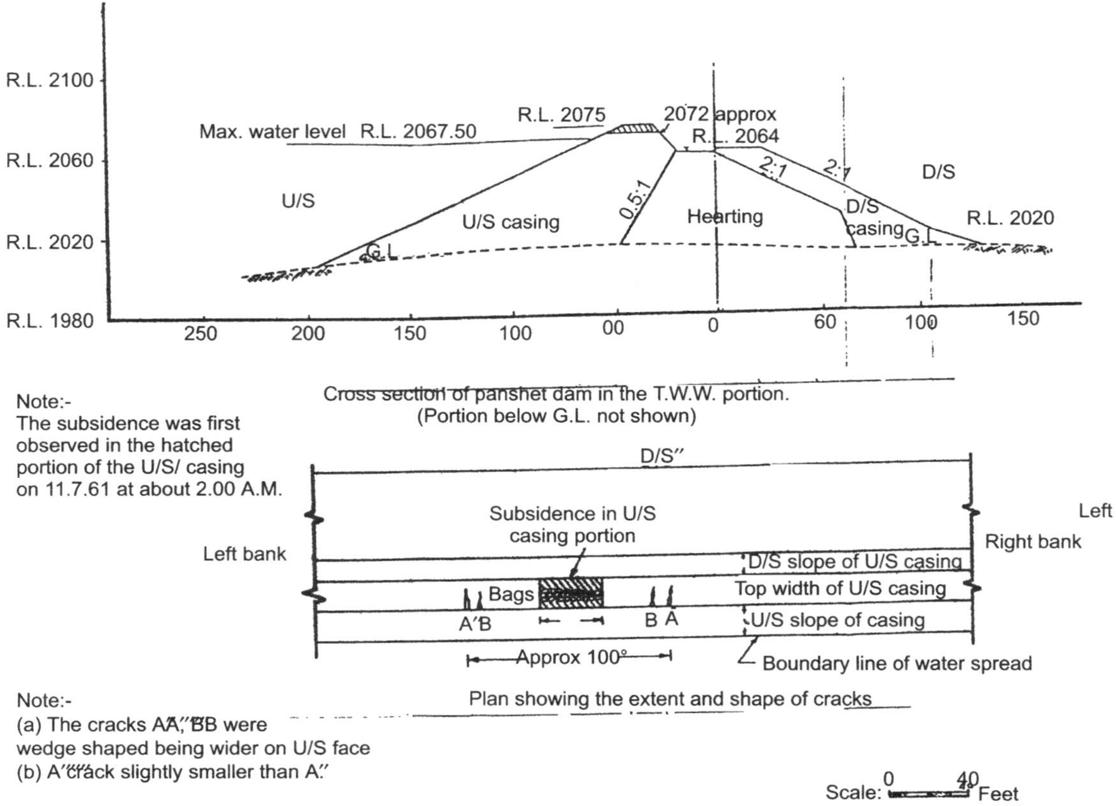

Fig. 4.25 Drawing showing position of cracks in T.W.W. portion observed on 11-7-61

theory of failure or any obvious explanation of the failure. There was every reason to believe that the failure had destroyed the most significant evidence to find out the real causes of failure and this posed a great challenge to the engineers involved in the investigations as also the Commission of Inquiry. The chairman of the Commission proceeded with an open mind, proposing that the engineers may provide hypotheses consistent with observations so as to be accepted as theory, explaining the causes of failure. It was also accepted that any one theory might not be able to provide explanation of all phenomenon, which may or may not be inter connected.

All the engineers agreed that the dam had not failed due to any of the following reasons:

- Inadequacy of spillway.
- Inadequacy of free-board.
- Heavy inflow for which dam was not designed.

It was also agreed that dam had not failed either due to wave-action or due to foundation failure.

The failure hypotheses were based mainly on following reasons:

- Internal erosion
- Cavitation vibrations
- Structural failures
- Piping.

If piping is considered as a part of the internal erosion, then only three factors remain. Opinions differed widely about these three hypothesis also, but most of the engineers agreed that downstream portion of the conduit must have been washed away after the dam was overtopped:

Internal Erosion

The Panshet outlet was designed to initially discharge 28.32cu.m.per sec. under a head of 41.75m in the first stage, assuming flow under high heads would be regulated through partial gate openings and conditions inside the conduit are provided appropriate gradients. The conditions existing in the conduit, before the failure, were quite unfavourable to discharge higher flows than the designed, under a fixed opening of 60cm provided by the outlet gate suspended by a wire rope. The hydraulic calculations based on flow depth at the end of the conduit corresponded to discharge of about 42.5cu.m. per sec., which was apparently improbable and, consequently, caused damages to the structure which must have been subjected to considerable vibrations. The high discharge must have caused pressure of the order of 3.3m for an average velocity in the conduit. The pressurized conduit drawing air through a free flow channel must have created "water hammer" conditions at the downstream end.

This was perhaps due to a break of the water column in the pressurized conduit and its rejoining when negative pressure wave developed, due to a sudden decrease of discharge in the conduit. The dynamic forces thus created most probably peeled out the mortar from joints of a voussoir, resulting in its dislocation and thus creating a gap. Erosion working backward from such a gap normally advances towards upstream leaving a cavity in the hearting zone of the earth dam cross section, which cannot be expected in relatively pervious zones due to non-cohesive material. Ultimately, the cavities collapsed one after the other, resulting in subsidence of the slope.

This hypothesis of failure did not take into account the leakages at the rock toe, which were perhaps considered of significance, as these were not considered as the cause of failure. It also provides no explanation to the washing out of the conduit pipe on the downstream. The Commission thus did not consider it as a convincing hypothesis.

The other argument put forward considered the phenomenon of the barrel running full, the condition for which it was not designed. This gave hissing noise and thuds. Secondly, significance was also attached to the flow of 0.14 to 0.28 cu.m.per sec., which was noticed at the rock toe.

The model studies carried out at the Central Water Power Research Station showed that when the head was above 30m, a standing wave was formed inside the barrel which was running full intermittently. The increase in head made the flow more turbulent and internal pressure varied from maximum of 6m to small negative pressures. These model studies, when projected to prototype, gave an estimation of higher fluctuating pressures, causing a shaking of the arch from within, resulting in continuous hammering from 7 July 1961, the day when the conduit was running full. Thus, the entire system was subjected to high frequency vibrations causing reduction in the strength of mortar at the joints and its disintegration, allowing a surge of water from the joints, dislocating the asphalt layer and, ultimately, the clay fill. These conditions must have been existing along the downstream half of the conduit wherever the joints opened.

The supposition that the mortar joints of the voussoirs were disturbed due to fluctuating flows that occurred along with high frequency vibrations, was not convincing to the other engineers. They did not agree with the magnitude of the pressure assumed to have caused the disintegration of the mortar. Even the experts and protagonists of the theory of cavitation vibrations differed with the theory and attributed the disintegration of mortar near the downstream end of the conduit due to fluctuating pressure at the cross-section, which had less over burden. The downstream end arch joints, which were under the strain of alternating compression and tension waves, got loosened resulting in disintegration of mortar and progressive failure. Those who did not agree with this argument firmly believed that the pressure in the conduit was a function of: (a) the flow as a pipe; and (b) the internal hammering, causing a high pressure. The pressure may be either due to wave, all along the conduit, or hammering at a particular point. The wave cycle itself was perhaps enough to cause loss in strength of the mortar. The zone involved would have extended to considerable length of the pipe between the wave front and the exit end, the fluctuation pressure being maximum in the downstream half of the barrel. All the inferences rallied around the comparison between the model and the projected prototype behaviour. Certain engineer experts maintained that the pressure should have been at least 4 to 5 times the pressure of the model and this was based on the reasons: (a) The models did not produce the cavitation phenomenon; (b) Many of the bubbles might have collapsed in the body of flowing water without hitting the surface; and (c) finally, the model was not dynamically similar to prototype and, consequently, the force with which the water would have hit the prototype must have been much more than experienced in the model. The values of pressure in the prototype, when compared to the values observed in the model, remained a matter of speculation. It was, therefore, reasonable to assume that the pressure changes along with variations in cavitation were sufficient to cause disintegration of mortar of the joints of the arches. The standing portion of the arch on the upstream proved the point. If the application of mortar in the joints of the voussoirs had been proper, the pressure of the order of 3 to 5 m would not have caused any opening of the joints and disintegration of the mortar.

Cavitation Vibration

According to an expert engineer on cavitation, cavitations did prevail in the region of the gate where negative pressure developed. This caused damage to guide rollers, and the retaining wall constructed on the transition box, which was found tilted on the upstream and fractured on the downstream. Severe cavitation erosion was noted at the base of the gate slots and the top horizontal seal appeared to have blown off. In the model, the vibrations in a high velocity flow were as a result of entrained air and these might have been more so in the prototype. The hissing noise or thud was also attributed to cavitation phenomenon. In more than one way, it was proved that cavitation conditions existed in the prototype and due to un-concreted bed of the conduit, disturbed flows and higher discharge than estimated, a highly pulsating flow was experienced by the structure. The cavitation damages, according to the engineer, always occur on the downstream in the high-pressure zone where bubbles collapse. Moreover, cavitation damages depend upon the strength of material and the period for which cavitation conditions prevail. The cavitation conditions might have been of low order but the duration at Panchet dam was considerable, namely 28 hours. The loosely packed rock toe also could not remain intact under sustained cavitation conditions and the watchman noticing the rock toe stones rolling down also provided evidence to sustain this theory. The model gave the indications of the defects in the design, the nature of vibration, pressure, flow conditions, etc., that existed in the prototype. The model, however, did not represent the cavitation conditions, as the study required different arrangements. Furthermore, the pulsating flow and the pressures varying from 0 to 6m in the prototype must have caused fatigue of the material and were the source of trouble rather than the steady flow. The loosened mortar must have been disturbed, exposing the joints through which water must have entered as the conduit was running full, and emerged at the downstream rock toe with a discharge of the order of 0.14 to 0.28 cu.m.per sec.. The engineer asserted that the cavitation conditions and consequent vibrations caused the failure of the upstream rock toe and subsidence of upstream pervious zone of earth dam leading to the overtopping of the dam. It was also presumed that the vibrations of the three elements, namely the gates, the tower and the conduit might have synchronized, causing dislocation of the rock toe. The frequency of vibrations based on the model observations was estimated to be of the order of 5 to 6 cycles per second. This might not have been quantitatively very close to the reality, but the trend did exist.

Other engineers, though in agreement with cavitation vibrations, did not agree that they caused disturbance of the rock toe. The cavitation vibrations of high order under high heads would have initiated large-scale sloughing of the slope. It was difficult to give any reason for the subsidence observed exactly overhead of the conduit extending to 10m length. This might have been due to transmission of vibrations to the soil slopes, which moved slowly, and finally, allowing the dam to be overtopped. The other explanation offered for the flow was that the joints opened due to loss of mortar under wave actions and jetting out of water from these joints.

The movement of soil under dynamic loading was not very well understood during those days and from observational experience and inferences, the cracks that appeared on the upstream and downstream slopes were attributed directly to the existing vibratory conditions as described above. The vibrations in non-cohesive soils in the presence of water caused a loss in the shear strength of the material and ultimately sloughing of the slope. This remained a disputed matter as some of the engineers were of the view that the cavitation vibrations created a near earthquake conditions whereas others did not agree due to a difference in intensity. The sustained vibrations for about 28 hours at Panshet, though of low intensity, did cause considerable damages.

Structural Failure

The Institution of Engineers (India), Pune and Bombay Chapters, had set up a committee to find reasons for the failure of dam due to the structural failure of the arch of the barrel.

One of the important reasons considered was the inadequacy of the thickness of the arch so as to sustain the loads, which were likely to come on it and particularly of the hearting portion of the dam. The dam height in this portion was about 45m and besides the pressure due to dam, there was additional pressure due to the drag of the soil surrounding the arch. The resultant thrust on the arch was not within the middle third of the joint and it is presumed that the high tension thus developed resulted in opening of the joints. The Committee was also of the view that heavy voussoirs arch, like that of the barrel, was not washed away bit by bit. When the dam was overtopped, the washed dam material, carried away by the flowing water, reaching the conduit end and choked the pipe. This resulted in an uplift pressure on arch from below and when this equalled the head upstream of the gate, the arch was lifted and fell into pieces. The upstream portion of the barrel counter balanced the uplift pressure and, thus, was relatively safe. All this must have happened in a very short time and the conduit continued to allow water through it after its failure. The rock surrounding the barrel was also found not very sound and had fissures to allow seepage of water and the committee attributed this as the possible piping source through the dam but did consider it to be significant enough to contribute to the failure.

The controversial subject was the load that must have come on the arch. A simple approach considered a load equal to the vertical load of the material above the conduit plus earth pressures that developed as horizontal load. There was a difference of opinion about the classification of the Panshet conduit with respect to its condition. It did not satisfy the condition of 'ditch conduit', as there was about 20m of filling above ground level. The fill above and below the ground was perfectly compacted except for the fill on the sides, in the narrow strip on the rock side, which might have been hand-tamped. The fissures in the rock carrying seepage water might have loosened the hand-tamped soil also. The cracks that were noticed on 11 July 1961 on upstream and downstream slopes—an expert engineer assumed—must have been continuous, passing through the hearting zone across the cross section. The portion of the dam—about 30m in length—in between the cracks was separated from the rest of the dam and was resting on the arch and the rock on the sides. The weight of vertical column of soil, about 35m in height, was worked out as 70.77 tonnes per square metre, which was directly acting on the arch. If settlement of soil prisms on the sides is considered more than the central prism, the weight will be of the order of 106.17 t/m^2 against the designed load of 94.69t/m^2 on the arch of the conduit. The Panshet conduit, if considered as rigid, perhaps did not get help from lateral earth pressure as it was not expected to yield as much as a flexible conduit. Moreover, the conduit resting in solid rock, an unyielding material was thus incapable of exerting active pressures.

There was another controversy about the consideration of arch of the conduit as a fixed or hinged arch. Apparently, it was a fixed arch and the design was based on this feature. It was argued that in hydraulic structures, arches should not be subjected to tension so as to avoid cracks. The matter requiring attention was the fact that the conduit arch failed at a section where the dam was not the highest. The arch failed about 6m upstream of the centre line of the dam where the height of embankment was about 35m, while its maximum height was 38m. A simple and straightforward reply was perhaps the poorest quality of earthwork in the embankment, representing 'the weakest link of a chain'. The difference in weight between the two sections, the highest and the failed one, was about 10%.

The committee of the Institution of Engineers gave a separate account of each of the three stages of failure, namely, subsidence, flow at the downstream end of the conduit and the failure of the downstream part of the conduit. The subsidence, which caused overtopping and consequent failure, is in no way connected with the other two events. The rainwater and the reservoir water passing through the fissures of the rock and weak zones of the side filling, taking the path of least resistance, collected at the downstream end of the conduit. The loose material and the debris, depositing at the downstream end of the conduit choked the conduit, resulting in blocking of the open end and creating the uplift conditions and ultimately washing away a portion of the conduit.

It was felt that the observed phenomenon, the washing away of the downstream portion of conduit, did not correlate with the theory of failure of structure. The washing away of a portion of conduit could not be taken as the proof of failure of structure. The assumption that the weakest portion of the arch was at the section where the conduit on the downstream has washed away might also not have been true. The arch stood the same loads without any sign of failure for about two months. The protagonist of the theory could not justify this cause of failure. There were differences in opinion about the lateral forces acting on the arch. The one argument that the lateral forces acting on the arch to be considered as zero did not hold well. The filling of soils with high moisture content between the conduit and trench was considerably thick and rose a full bank above the conduit. It was, thus, capable of exerting horizontal pressures. The cut of the Panshet conduit was about 24m deep and height of the conduit was about 4.5 m but was not considered as 'ditch conduit' as the fill above the crown of the conduit was about 20 m and this was not loose but highly compacted. This evidence did not satisfy the Spangler's criterion of 'ditch conduit'. Hence, engineers were tempted to calculate loads as in the case of culvert. This controversy did not solve any problem to justify the arch failure. The concrete experts also felt that in place of soil if the gap between the conduit and the rock was filled with concrete, besides being expensive, it would not have been more effective. Spangler—specifically considering the Panshet conduit—was of the opinion that it was a case of 'ditch conduit' and lateral pressures remain effective for underground conduits irrespective of their flexible or rigid condition.

According to Spangler, the magnitude of lateral pressures would not have been more that 0.2 to 0.25, depending upon the clay used for filling.

Considering the above arguments and the failure that occurred, it was concluded that the arch had not collapsed under the embankment load. Under excess load, the arch voussoirs would have readjusted causing a reduction in the vertical load and an increase in the horizontal passive load. There was no doubt about the joints opening at high-tension points and mortar crushing under high compressive stress.

Besides the various theories presented at the Commission of Inquiry to justify different reasons responsible for failure of Panshet dam, there were some altogether different interpretation given to sequence of the failure phenomenon.

Murti (1977), based on his own experiences with failure of embankments, considered the following phenomenon to be of great significance in the case if Panshet dam:

1. Closing the gap of about 41 m height in a short interval of five weeks, followed by heavy rainfall, resulting in fast filling of reservoir caused settlements during construction and the first filling.
2. Close to the steep sides of the rock trench above the conduit, there were differential settlements due to great difference in height of embankment. The height of bank in the trench was about 41 m, that is double of the height of bank on either edges of the trench.
3. There was no protection or provision of a cutoff wall or rib to stop the seepage or leakage of water along the interface.

Lack of coordination among various experts was very obvious, as had been brought out consciously or unconsciously by the experts during their presentation before the Commission. The expert engineer in one field overlooked the significance of other aspects falling outside his own interest. Perhaps this might have been the case at the design stage also. However, it is obvious that none of the above causes could have individually caused the failure of the two dams.

KHADAKWASLA DAM

As mentioned above, the Khadakwasla masonry dam was located about 16km from the city of Pune, on the downstream of Panshet dam. This dam breached on 12 July 1961 at about 14.30 hours. The breach was caused as the flood water from the breach of Panshet dam started gushing into the Khadakwasla reservoir and the water level standing at 10m levels rose to about 13 m in a short interval of one and a half hours.

Khadakwasla dam is a straight gravity dam with its cross-section in Fig. 4.26. The dam was built in rubble masonry in lime mortar without drainage gallery, considering the rock to be impervious. Uplift forces were not considered in design of the section. The dam was 80 years old when it breached.

When Panshet dam construction was taken up, in the background of hydrological experiences faced by Khadakwasla dam, there was a serious thinking of strengthening it. It was felt necessary to improve its stability under higher water levels by considering leaching of lime, leakages and most important being uplift conditions. Before any action could be taken, the completion and failure of Panshet dam resulted in the breach of the Khadakwasla dam. At the time of failure, all the gates were open and water overtopped the entire length of dam and rose to a height of 2.7 m above the top of the dam. The overflow, causing scouring of soils on the downstream, exposed the lean section of the dam under the water level of 4 m above the designed level. The section withstood the dynamic forces due to overflow for about four hours before it breached. The breached section did not happen to be the highest. The dam failed at the section where there was a step in the foundations, consisting of typical geology of area, Deccan trap, i.e. basic basalt formations belonging to Aryan group.

Figure 4.27 shows the plan, longitudinal section and the cross-section of the dam after repairs.

Causes of Failure

When the Panshet dam, located about 22km upstream of Kadakwasla dam failed, it generated a huge load wave. The water released due to the failure of the Panshet dam was estimated as about 200 Mcu.m. This flood wave impinged on the old Khadakwasla lake which was already full with

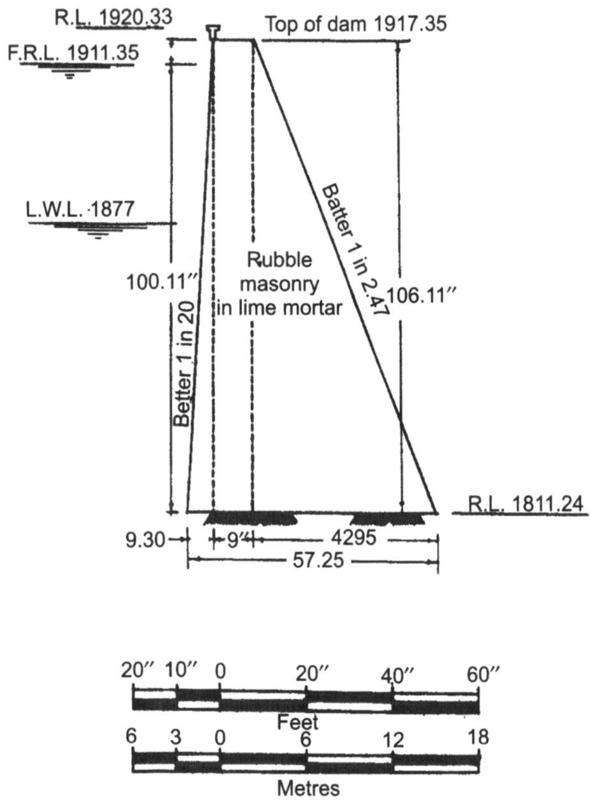

Fig. 4.26 Cross-section of Khadakwasla dam at ch. 900

all the gates nearly fully open. The onslaught of the rushing water was so great that the water actually overtopped the entire length of the dam and rose to a height of 2.75 m above the top of the dam. Observers reported having felt the dam vibrating on that particular day. The overflow scoured the foundations on the downstream side. Yet this dam withstood a water level 4.25 m above the design level and after withstanding the dynamic forces due to impact of flood wave for nearly four hours, it failed at a section where there was a step in the foundation. The failure took place when the over flood was about 1 m above the dam.

Repairs

In the year 1964, it was decided to repair the Khadakwasla dam. The repair was carried out by first filling the gap with masonry, followed by

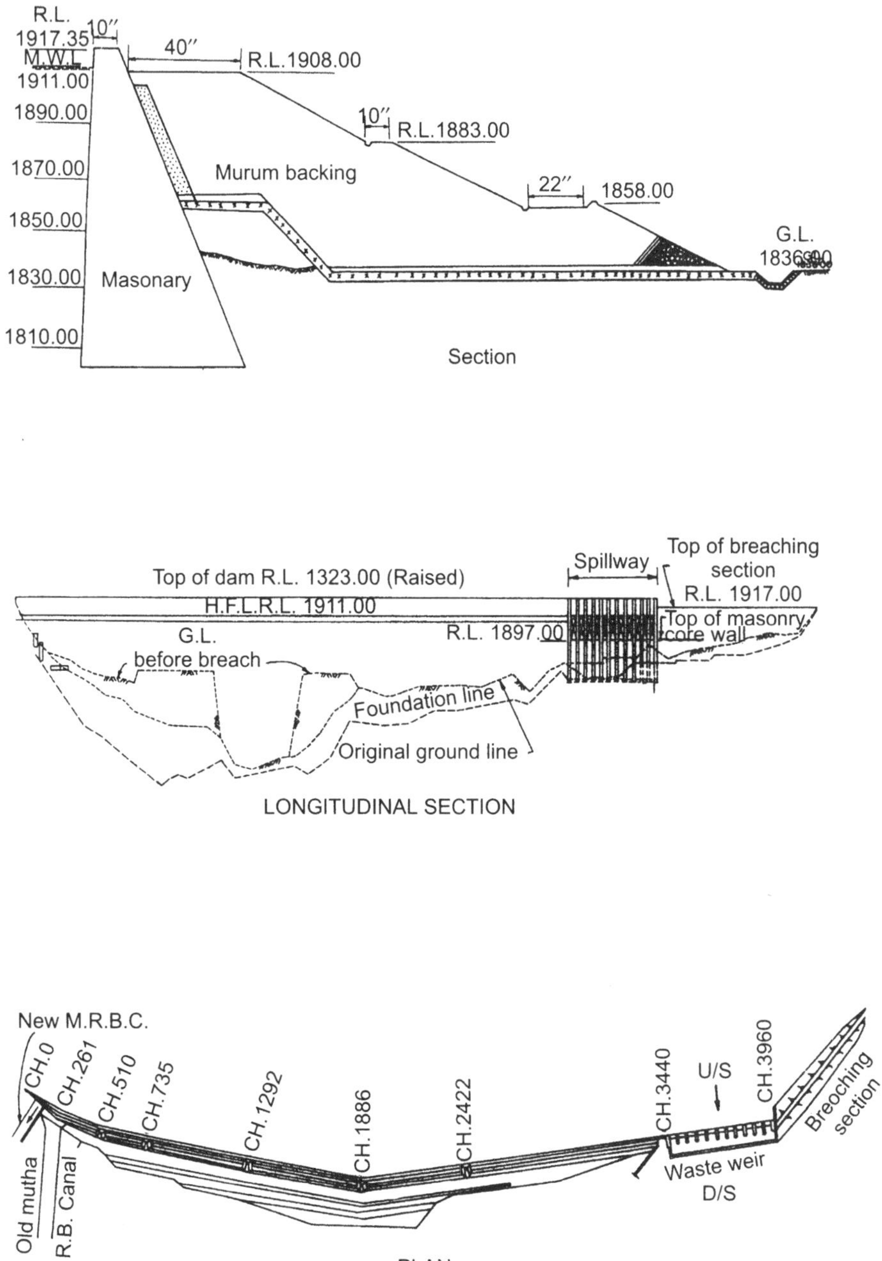

Fig. 4.27 Plan and Sections of Repaired Khadakwasla dam

providing a pervious earth backing with a sand filter on the downstream face of the masonry dam. An adequately designed radial gates spillway was provided. Further, a shallow breachable earthen fuse plug was provide which could be overtopped or breached in the unlikely event of another flood. To increase the vertical weight and increase the free board, the height of the masonry portion was increased by 1.83 m.

There was unanimity among the engineers that the failure of Panshet dam in the early hours of 12 July 1961, caused unprecedented flood conditions and overflowing of the Khadakwasla dam. The dam was thus exposed to stresses for which it was not designed, resulting in breach.

BALDWIN HILLS RESERVOIR

Introduction

This failure is one of the four major dam failures which was re-examined at an International Workshop on Dam Failures, Lafayette, Indiana, USA in August 1985, about two decades after the failure of Baldwin Hills in 1963. Immediately following the failure, a number of independent investigations were carried out, but none of the reports—it is believed—was made public or published for some reason or the other. However, certain experts involved at some stage of investigation, planning and construction, (Jansen, Dukleth et al., 1967; Leps 1972; Casagrande, Wilson et al., 1972), presented their observations and analyses, besides the Proceedings of the International Workshop (1985). One of the authors, Dr Penman happened to be closely associated of Water Resources, California (1964), on the investigation of failure of the reservoir.

The Baldwin Hills dam, constructed during 1947–1951 at Southwest edge of Los Angeles basin, near Beverly Hills and Newport Beach, California, USA, failed on 14 December 1963. About 71 m high, this homogeneous earthfill dam with upstream lining was 198 m long. The main dam on the northern side had compacted earth dykes on the others. At about 11.15 am, on the day of the failure, as the water from the drains could be heard under the reservoir lining, the storage level was reduced. This would have taken 24 hours to empty the reservoir. At about 1 pm, muddy leakage from the east abutment of the main dam was noticed and reduction in the water level revealed a meter break in the reservoir lining, opposite the downstream discharge. Nothing could control the water discharge that increased and breached the downstream face, emptying the reservoir by 5 pm. After the reservoir was first filled in 1951, it is on the day of the failure in 1963 that the collapsed lining exposed a large voids, perhaps formed by continuous piping for a long time. In four and a half hours—the time that took the reservoir to get empty—thousands of people were evacuated but due to sudden rush of water, property worth millions of dollars was lost.

Reservoir Site, Design and Construction

The site of the Baldwin Hills reservoir, shown in plan Fig. 4.28, had a series of low hills, extending in northwest-trending line along the southwestern margin of the Los Angeles Basin. The geology was highlighted during surface mapping and extensive subsoil exploration for petroleum. The reservoir hills site, in general, is a part of Newport-Inglewood uplift, intercepted by a series of faulted anticlinal dome formed by oil and gas tapped from the wells in the oil-filed area. The hills are made up of sedimentary formations of marine origin, overlying a crystalline schist at 3000m depth. The sedimentary formations in the foundations of the reservoir mainly consist of silt, sand and gravel. Lateral and vertical displacements in the faulted zone in the vicinity of the area—besides warping of the beds—exist. Further displacement, if

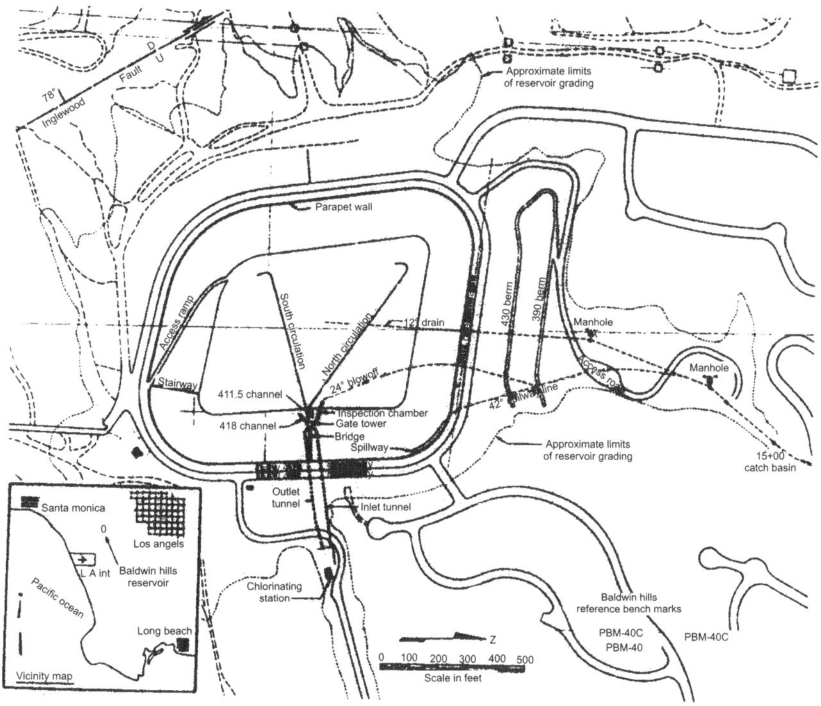

Fig. 4.28 Site plan of Baldwin Hills reservoir. Casagrande, Willson and Schwantes (1972)

any, were not considered as a threat to the safety of the reservoir. However, in certain fault zones, flexibility was provided in the design. In view of foundation soils, under drainage and impervious lining of reservoir appeared imperative. The stability of embankment was taken care of by ensuring that seepage water does not find its way to the foundation soils. An impervious reservoir lining as shown in Fig. 4.29 was provided. Two under drainage systems were designed; one to drain the foundations under the earth embankments and the other, reservoir under drainage, to drain the seepage passing through earth lining, as shown in Fig. 4.30. The seepage was conveyed to a measurement chamber. Under the main embankment, a 30cm foundation drain of clay tiles was placed with the upper half covered with a lean cement mix of pea gravel to allow for water seepage. The composite reservoir lining consisted of 7.5cm thick porous asphaltic layer laid over compacted earth lining, varying in thickness from 3m at the bottom to 1.5m normal to the slope at the top. Below the earth layer, a 10cm thick drain blanket of pea gravel of lean cement mix was laid. The top of this was sealed with a 6.5 mm porous gunited sand layer to prevent the entry of soil particles. The lowest layer of lining had an asphaltic membrane, about 6 mm thick laid over natural or compacted fill. Cotton fabric was used to reinforce the membrane wherever necessary.

Seepage problems during construction of foundations of the main dam were tackled by providing flow of water through horizontal drains, pea

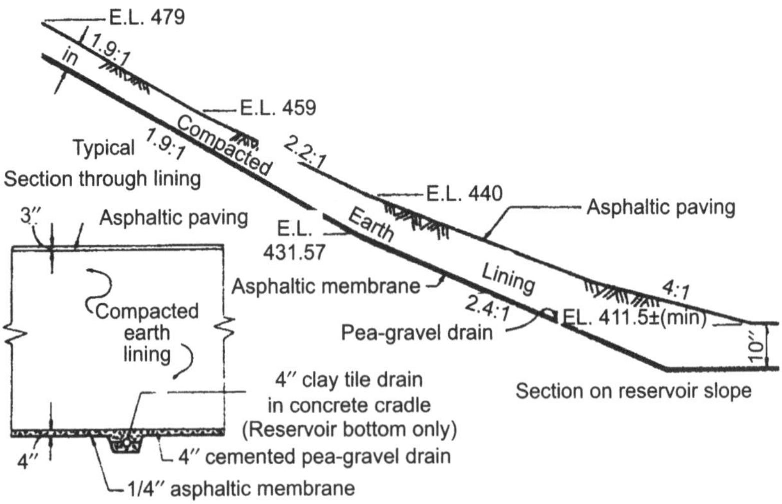

Fig. 4.29 Cross-section of reservoir lining. Jansen, Dukleth, Gordon, James, and Shields (1967)

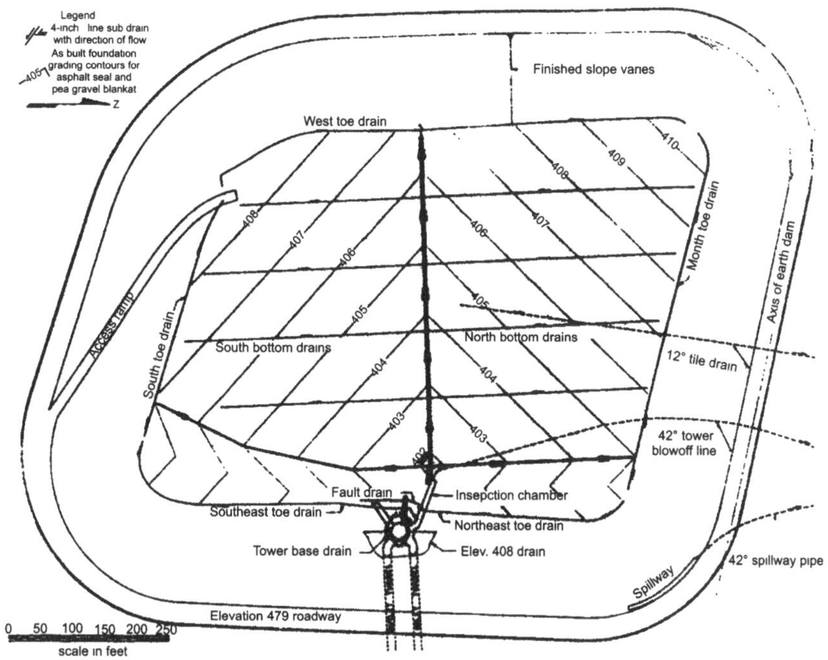

Fig. 4.30 Reservoir floor subdrain system. Casagrande, Wilson and Schwantes (1972)

gravel blanket, etc. Slides were encountered while stripping the abutment mainly in clay beds with movements of the order of 10 cm along the failure plane. Stabilisation of these required horizontal and vertical drains. The surface cracks that appeared were excavated to shallow depths and backfilled with soils, ignoring the actual depths of cracks. Contribution of these movements and cracks to failure, directly or indirectly, is not ruled out. It is difficult to consider the lining as truly watertight, due to the horizontal cracks that appeared in the asphalt membrane during placement. This treatment of the subgrade soils before lining was taken up might not have been as uniform as it was expected to be and hence the possibility of differential settlement of pea-gravel blanket is not ruled out. In such a situation, there was the possibility of seepage water travelling down to the foundations.

Operational Problems and Maintenance

During the filling of the reservoir for the first time in April-May 1951, heavy leakage of the order of 284 1/min., through under drains developed, forcing the maintenance crew to drain the reservoir water for repairs. The dye tests conducted revealed the collapse of asphaltic lining at the toe and compression cracks. Considerable maintenance was needed for the drainage system under the reservoir during the early years of operation. Treatment was necessary to prevent plugging of drains due to asphalt that flowed from the membrane and formation of calcium carbonate deposits in the drains. In the first year of operation, a crack was noted in the drainage inspection chamber, which continued to widen over the following years of operation. These happenings alerted the owner of the reservoir, the Department of Water and Power, Los Angeles, California, who continued to record surveillance data by periodic inspection and through measurement of strain and earth movement. Earthquake recorders and tiltmeters on east and west sides of the reservoir as also on the main dam were also installed. The water levels in the observation wells along the perimeter of the reservoir were also recorded to check the water in the foundation or embankment. The extensive collection of surveillance data was necessary in view of the complex geologic features of the site and highly inhabited surrounding areas.

The surveying—carried out over a long period of the Baldwin Hills area, from 1910 to 1963—showed that subsidence of ground surface had been more than 3.5 m. Baldwin Hills Reservoir site is within the limits of the subsidence bowl. Consequently, the foundation strata beneath the reservoir must have progressively stretched horizontally. This movement had impact on the faults in the area, causing physical separation of certain areas into blocks. This happens to be in the vicinity of the most productive zone of the Inglewood Oil Field, commissioned in 1924. The extraction of oil was followed by field repressurization projects by injecting brine solution in 1954, 1960 and 1962. Initial subsidence in the Baldwin Hills area coincided with the first extraction from the oil field.

Seismological laboratory records, from a region about 25km from the reservoir site, revealed that on the failure day and even during the preceding month, the reservoir area did not experience any earthquake shock large enough to cause damage. There is the possibility of rupture in the tile drain due to earthquakes in the past or any creep that have been taking place in faults.

The internal compression of the embankment and compression of the sedimentary foundation soils under the load of the dam was an expected phenomenon and these were recorded to be within normal limits. Maximum settlement of the reservoir in the north-south direction, indicating that foundation failure was in progress along certain faults. Before the failure of the reservoir, the vertical settlement at one of the faults increased—an followed by uplift at the adjacent structures. A similar uplift was noted during operation of the reservoir, but of a lesser magnitude. Separation of various concrete structures—adjacent and around the reservoir—parapet walls and construction joints, were in progress from a long time but gained acceleration near about the failure of the reservoir.

Investigation of Failure

As stated above, numerous investigation teams were formed by various agencies with the specific purpose of investigating the causes of failure of the reservoir on Saturday; 14 December 1963, in 4 hours and 25 min., after the first warning at 11.15 hours. The investigations carried out by the Department of Water Resources, State of California were for the first time well documented in a report published in April 1964. This report formed the basis of further subsequent investigations. Arthur Casagrande, Wilson and Schwantes (1972), the participated in thorough investigations from 1964 to 1971. The conclusions drawn by various investigators and reviewers varied considerably, immediately after the disaster and even when the failure was re-examined in 1985. It was difficult to connect the obvious failed elements to pinpoint any flaw in the design, except that there was a lack of redundancy and the safety depended only on efficient working of the drainage system and intact functioning of the reservoir lining made up of composite membranes. Leps (1985), it is of interest to note, added in the International Workshop, that during the construction, the designer had invited Dr Terzaghi to the site for inspection of the dam, the under drainage system, the instrumentation, etc., but he refused to offer any comment.

Causes of Failure

The Department of Water Resources, California (1964) was of the view that movements of faults, spread over the Baldwin Hills area, caused progressive stress and displacements of foundations. This caused a rupture of the reservoir lining and flow of water under pressure to the foundations and faults. This crack further resulted in erosion of foundations and uncontrolled leakage and, finally, the total failure. Jenson, Duklet, et al., (1967) considered progressive subsidence, that was observed in the vicinity, as the reason of earth movement, a phenomenon existing even before the construction of reservoir. According to them, a well-designed strong dam satisfying the safety criterion, to store water was unsuitable for the geologic formations of the site. According to Casagrande (1965), the risk taken by the designer, it is difficult to decide, whether it was a 'calculated risk' or the 'underestimation of risk'?

Leps (1972) was of the view that the under drainage system was subjected to progressive strains and vertical offsets of the foundations at the crossings of fault. This resulted in collapse of the lining and unabated flow of water to foundations till the reservoir was emptied. He also felt that the seepage of water through the faults may not be more than few litres per minute, but this continued for a long time, eroding the local material and forming narrow and long cavities. These cavities may not have initiated the failure but definitely must have accelerated the outflow of water from the reservoir.

Casagrande (1972), who had taken up investigation for Standard Oil Company of California, also concurred with the conclusions drawn by others, that the reservoir failed mainly due to erodible foundation strata and the faults of different magnitude crossing it. The main approach of the designer in providing impervious lining membranes and the drainage system was to avoid water ingress into the dam and its foundations. If any seepage takes place, the drains would convey the water away from the dam foundations. According to him, the greatest contribution to failure was by the brittle tile drain, which collapsed under strain and hastened the failure. Failure of the reservoir was unavoidable due to the site conditions, but if the drain was not provided, perhaps the erosion of foundation soils would have taken a longer time. Piping is one critical phenomenon in safety of a dam, which is difficult to monitor, unless it highlights at the open side. At this stage, the situation becomes difficult to manage and many a times impossible, as there is no inbuilt factor of safety for it in the design of embankment. The Baldwin Hill area, including the reservoir site, was experiencing subsidence, before and after reservoir completion. It remained a controversial issue that how much movement was caused by oil recovery or fluid injection.

Hamilton and Meehan (1971) were of the view that injection of fluid had caused shear displacement, which had travelled up from the fault to the lining and this triggered the failure. It was not convincing as vertical displacement was not observed along the faults anywhere in the area during investigations, as was observed by Casagrande.

Osterberg (1989) in the 21st Terzaghi lecture, stated that 'redundancy' in design, i.e. if one design element fails, the other element can take over the function, and in doing so, can save a catastrophic failure and the damage is limited. This amounts to provision of a second line of defence in design, in the event of failure. In case of leakages, self-sealing arrangements in the design of dam can help in controlling seepage. In the case of Baldwin Hills Reservoir, the entire safety of reservoir depended upon the water tightness of composite lining and functioning of the subdraiage system without getting affected by any movements of ground. These conditions could not be maintained due to prolonged seepage conditions and movements at the faults.

Scott (1987), while delivering the 27th Rankine lecture on 'Failures', has discussed in detail the failure of Baldwin Hills Reservoir. He attributed the failure to: (i) either a sudden geologic movements or long-term creep of faults; (ii) withdrawals of oil, water and gas and repressurization, causing down and up movements of ground; and (iii) rupture of the lining and drainage system due to greater water pressure on top of the lining, resulting in greater compaction on one side of the fault rather than on the other. He expressed the difficulty in the analyses of such a failure as a true picture of seepage and piping process remains based on speculation.

Leonards (1985), in conclusion of the Workshop, said that there was no way to ascertain or monitor, whether or not the impervious membrane was intact, which was the basic and only safety element in design of dam and reservoir. It is difficult to pin point any other reason.

VAIONT DAM

Introduction

Vaiont Dam, built in the Italian Alps in Northeastern Italy at 680m above sea level, is located 46.3 N and 12.3 E near the border of the Veneto and Friuli-Venezia Giulia regions. It is centrally located in the region, offering an ideal situation for hydroelectric power production. The canyon is steep-sided, narrow and glaciated, opening to populated lowlands to the west and is an ideal site for a narrow crested high dam and deep reservoir.

In this canyon, the 265m-high Vaiont dam is built on the Vaiont River, a tributary of the Piave River, in Venetia, Italy in the southeastern part of the Dolomite Region of the Italian Alps, near Belluno. It is located about 100 km north of Venice. The dam, a doubly curved thin concrete arch dam, said to be the world's highest thin arch dam and the second highest dam of any kind, was completed in 1960 and used to generate hydroelectric power for the rapidly expanding northern cities of Milan, Turin and Modena. The arch is 3–4m thick at the top and 22.7m thick at the bottom of the canyon. The chord of the dam was 160m and the volume of impounded water was 115 million m^3. After heavy rains in 1963, landslides into the Vaiont reservoir caused the stored water to spill over the dam, sweeping away the villages of Longarone, Pirago, Villanova, Rivalta and Fae. The Vaiont reservoir disaster is a classic example of the consequences of the failure of engineers and geologists to understand the nature of the problem they were trying to deal with. During the filling of the reservoir, a block of approximately 270 million m^3 detached from one wall and slid into lake at velocities of up to 30 m/sec. As a result, a wave over-topped the dam by 250 m and swept onto the valley below, resulting in the loss of about 2500 lives. Remarkably, the dam remained unbroken. The great reserve of strength inherent in an arch dam were thus dramatically displayed when the reservoir behind Vaiont dam was virtually destroyed by the landslide.

Geology

The geology of Vaiont dam site is characterized by massive, near-vertical cliffs formed in the Jurassic Dogger formation and underlying Triassic formations. The local valleys tend to be associated with outcrops of the weaker formations, particularly the Upper and Lower Cretaceous and Tertiary units, which contain more clays and are more thinly bedded. Thus, the generalised geological structure is of a syncline cut by the valley. The syncline is based in middle Jurassic limestone, overlain with successive layers of upper Jurassic limestone with clay and Cretaceous limestones.

The site of the Vaiont reservoir is on a deep sedimentary formation, composed primarily of limestone with interbedding of claystone and marl. This rock has low shear strength and the limestone has been subjected to solution by groundwater. This had led to extensive cavities and tubes and enlarged joints and bedding planes. The labyrinth of solution conduits reduced the integrity of the formation and facilitated a gradual increase in internal water pressures. This, in turn, diminished the friction between the rock interfaces.

Main Geologic Features:

(i) Steep canyon walls composed of interbedded limestone and shale; badly cracked and deformed structures; open fractures in the shale are inclined toward the reservoir body. The steepness of the canyon walls enhance the strong driving forces (gravitational) at work on rock structures.

(ii) High potential for bank storage (water absorption by canyon walls) into the groundwater system that will increase the water pressure on all the rocks in contact with the reservoir.

(iii) The nature of the shale beds is such that cohesion will be reduced as its clay minerals become saturated.

(iv) Evidence of ancient rockfalls and landslides on the northern side of the canyon.

(v) Evidence of creep activity along the southern side of the canyon.

Events during Construction and Filling

Large volumes of concrete were injected into the bedrock as "dental treatment" in an attempt to strengthen the fractured rock. During reservoir filling on 4 November 1960, when the reservoir water depth was about 130m, 700,000 cubic metres of rock and soil slid into the reservoir from the southern side. After the 1960 slide, the lake level was generally not allowed to exceed 680m, i.e. 42.2m below the crest level of 722.5m. As the valley filled with water after completion of the dam in 1960, an ancient landslide on its upper southern side, adjacent to the dam, began to move again, in episodes of slow creeping movement. It was later established that this was due to groundwater unable to escape into the floor of the now-flooded valley, saturating a layer of clay within the rocks beneath it. This accumulation of water high in the slope was most marked during periods of heavy rain, although the association of heavy rainfall and creep went unnoticed. However, the presence of the impermeable clay layer in the bedrock was also not recognised at the time and it was assumed that the movement was due to local saturation of the rocks below the level of water in the reservoir, at the toe of the creeping landslide, rather than accumulation of water pressure in the entire mountainside. It was, therefore, proposed to regulate the movement of the landslide, and allow it to settle to a new equilibrium, by lowering the level of water in the lake when an episode of creep was in progress, until the toe of the landslide was no longer saturated and the creeping stopped. The reservoir was then allowed to refill and the drainage cycle repeated whenever any creeping movement occurred again. It was observed that the creep rate increased with a rise in the level of the reservoir. The creep rate, which was about 1cm per week by January 1963, increased to more than 1 cm a day by mid September 1963. Animals grazing on the slopes the area abandoned the region on 1 October 1963, presumably sensing the impending heard. At the end of September 1963, the reservoir level had reached a maximum of 710m, but prior to the disaster, it was lowered again to 700m.

On 8 October 1963, after ten days of heavy rain, the dam engineers attempted to arrest the creeping motion of the landslide by partially draining the reservoir again. However, because lowering the level of the lake immediately removed a stabilizing force, that of the weight of water in the lake pushing against the toe of the landslides, whilst not immediately draining the water from the upper part of the mountain because of the barrier to flow represented by the clay layer. The net effect of their action was actually to destabilize the landslide rather than stabilize it. The movement accelerated through 9 October 1963, as the engineers continued to attempt to arrest the movement of the slide by lowering the reservoir level (although, in fact, their efforts merely served to remove the arresting buttress provided by the weight of water on the toe of the slide). Late that evening at 22.38 GMT, the moving rock mass, with a volume of over 240 million cubic metres, failed catastrophically and slid into the reservoir within about 30 seconds, at an average velocity of 20–30 metres per second, or upto 110 km/h, shaking seismographs all across Europe. The slide created strong earth tremors which were recorded in Brussels, over 500 miles away. The translational slide involved a 2km long, 1.8km wide and 200m thick slab of limestone. The mass was a prehistoric slide which had come down from a mountain located on the southern side of the reservoir. The instability of the slope of this mountain had been recognized by some of the project planners as a serious disadvantage of the reservoir location.

Remarkably, the dam withstood the weight of rock piling up behind it and remains largely intact to this day. No damage was caused to the mail shell or abutments even though the dam was hit by a total force of about 36,000 mega newtons from the slide and over-topping pressure. The debris filled the reservoir for about 2km upstream from the dam to depths of over 300m. As such, the dam can no longer be used because the cost of removing the slid material is too great. The landslide placed as much as 50% of the water within the lake: a water wave travelled eastwards up the reservoir, destroying two villages, whilst to the west, water surges of water swept up the valley sides to heights of 300m above the reservoir level and over the top of the dam. The spillway bridge was torn away, 70m high flood wave swept over the town of Longarone and both up and down the Piave valley. A total of about 2500 people were killed.

An analysis of the translational slide indicates that it was caused by a combination of factors including:

(i) Adverse geologic features in the reservoir area.
(ii) Manmade conditions imposed by impounding water with bank storage, affecting the otherwise delicately balanced stability of a steep rock slope.
(iii) Progressive weakening of the rock mass with time, accelerated by excessive groundwater recharge (excessive rain).

This disaster of the Vaiont dam sparked renewed attention to slope stability around the world. The walls of the valley of the reservoir were particularly susceptible to large landslides: (a) the slopes were steep (~28); (b) the rock layers dipped toward the valley floor; (c) the rock consisted of a highly fractured, clay-rich limestone with abundant dissolution features, allowing a well-connected hydrologic system to exist; and (d) the site was the scene of prior large slides in the geologic past. The possibility of a landside into the reservoir was recognized in advance but the estimates of its magnitude were underestimated by two and one-half orders of magnitude. Surveying measurements and water table (piezometric) measurements made *after* the reservoir began to be filled showed that changes in the reservoir level caused near simultaneous changes in the water table in the valley walls, and that increases in the elevation of the reservoir level corresponded to increased displacement rates of the slide mass on the south valley wall. The failure of the dam can be attributed to inadequate recognition, characterisation, and evaluation of the conditions of the slopes flanking the reservoir before and after the reservoir was developed.

How could this Disaster have been Avoided?

Numerous preventive measures could have been taken to avoid or decrease the likelihood of a disaster of this magnitude from happening again.

Terrain Investigation: Before any structure dependent on geology is designed, one should check if the location is suitable, investigate the terrain, and note the likelihood of plate movement and the types of rock substructure present.

Testing: Evaluate the strength of the materials, conducting proper field tests in which the environment of the valley can be simulated, providing more accurate and relevant statistics.

Design: Anticipate the trends that occur in the area, that is since landslides were known to happen, the dam should have been designed to anticipate a large-scale landslide, even though they were not a normal occurrence.

Evacuation Plan: This area had experienced landslide activity earlier also, so before completion of such a large potentially environmentally altering structure, the safety of the residents should have been considered to be a priority.

KOYNA DAM

Introduction

An earthquake of the magnitude of 6.5–7 (Richter scale) occurred on 11 December 1967 in the Koyna region, about 200 km to the south-southeast of Mumbai India. The quake took its toll of lives, reportedly about 180, besides injuring several hundred persons. Most of the buildings within about 30 km from the epicenter collapsed.

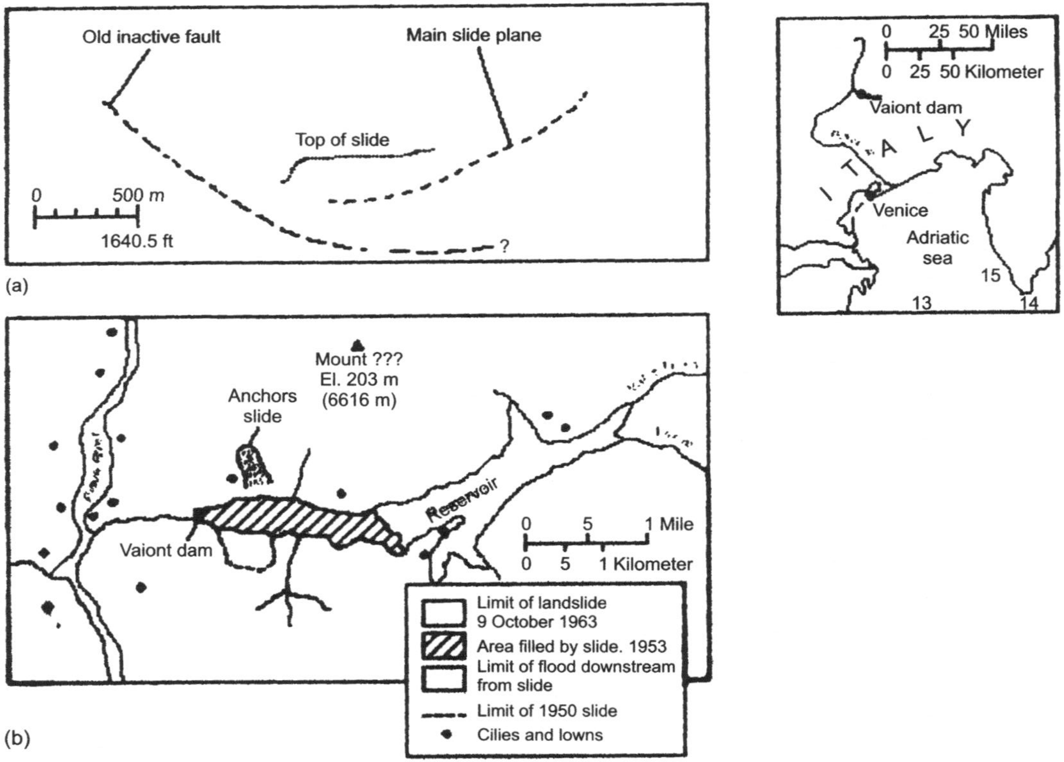

Fig. 4.31 Vaiont reservoir disaster, (a) Cross section of the Vaiont River Valley, (b) Map of the disaster site noted with evidence of previous activity, (after G.A. Kietsch, "The Vaiont Reservoir Disaster", California division of mines and geology. Mineral information service, vol. 18, no 7, pp. 129-138)

The Koyna Hydro-electric Project is built across the River Koyna, a tributary of the Krishna river. The project, with an installed capacity of 540 MW, was commissioned in 1962. When the earthquake occured in the region, in the year 1967, most of the component structures of the project stood well, though the dam suffered some damages.

Very rarely does a solid gravity dam like the Koyna Project suffer damages due to earthquake. Some details of the project, the geology, the earthquake, the damage, remedial measures and some aspects of Reservoir Induced seismisity are described below.

Location and Layout of the Dam

Koyna dam in India is one of the few concrete dams that have experienced a destructive earthquake. The dam was designed as a straight gravity dam made of rubble concrete. The dam is about 640m long and 104m high above the deepest foundation level. The dam is founded on basalt and consists of 46 independent monoliths, each having a width of 15.24m. Adjacent monoliths are separated by contraction joints provided with double water seal. The joints are ungrouted and the monoliths, thus, act independently. Figure 4.32 shows the Plan and the Elevation of the overflow and the non-overflow sections, while Fig. 4.33 is a line drawing of the cross-section. It would be seen that the cross-section of the non-overflow dam is rather unusual. The principal features of the project are given below:

I. Reservoir	
Catchment Area	: 2.775×10^6 TMC
Water Spread	: 115.35 km^2
II. Dam	
Height above river bed level	: 85.34m
Height above deepest foundation level	: 103.02m
Length at road level	: 853.44m
Flood gates (radial)	: 12.5×7.6m each
III. Intake and Head Race tunnel	
Intake channel, length	: 1134m
Head race tunnel length	: 6.4m diameter, circular, 3749m concrete lined
IV. Pressure Shafts	
Number of shafts	: 4, steel lined
Length with take offs	: 731.5m each
Diameter	: 2.59m at the lower end
V. Power House and Associated Structures	
Approach Tunnel length	: 6.7m in width, 864m in unlined but gunited

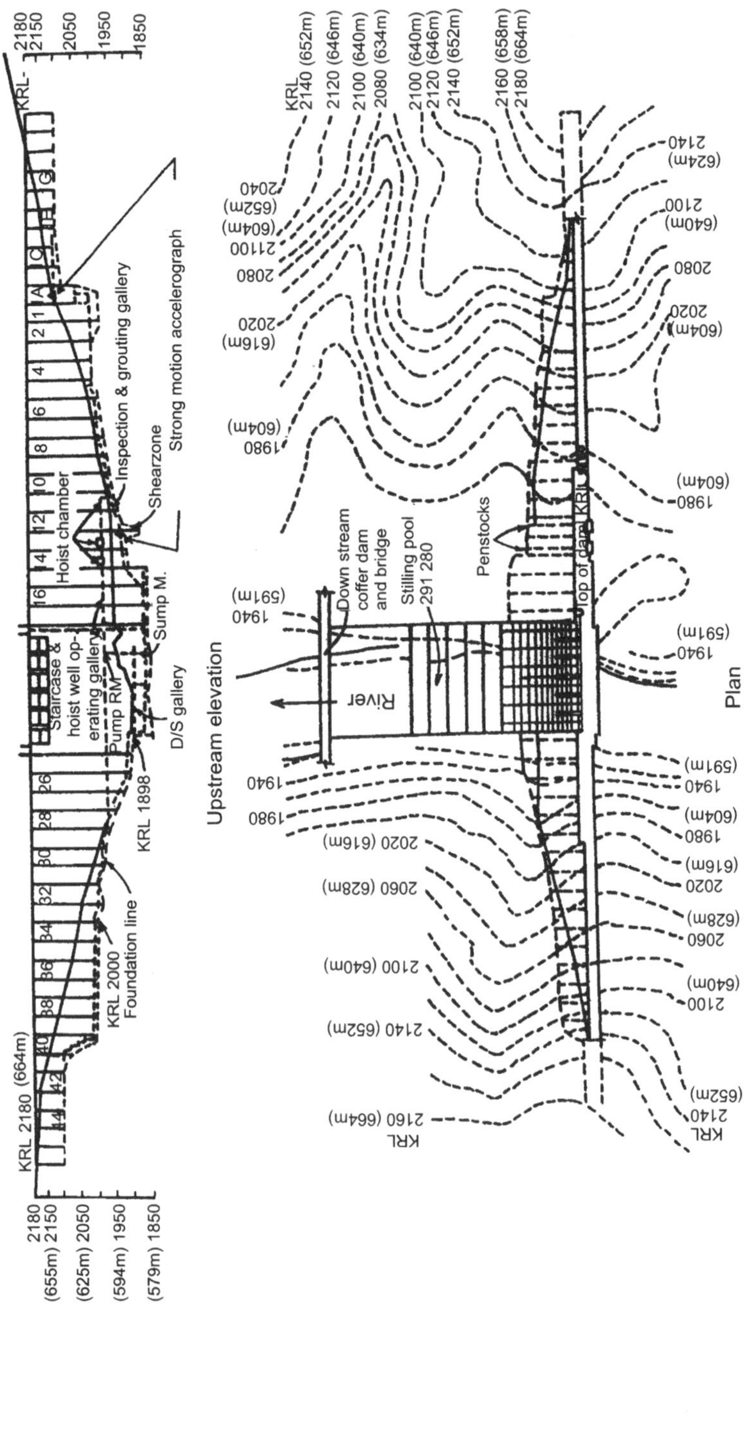

Fig. 4.32 Plan and elevation of Koyna dam

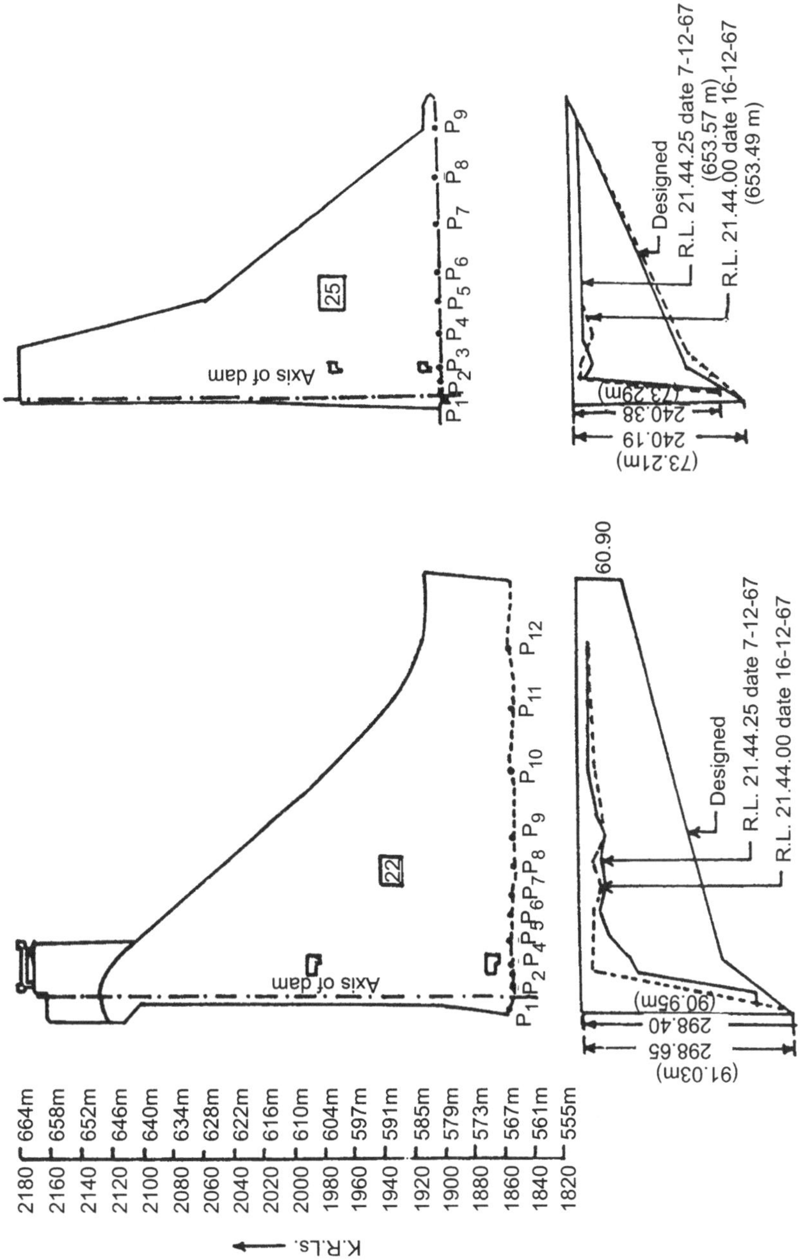

Fig. 4.33 Cross sections and uplift pressures of koyna dam

Underground Power House	
Machine hall	: 15.85m × 193.2m
Transformer hall	: 12.95m × 199.95m
Valve house	: 6.09m × 173.89m
Tail race tunnel	: Size: 7.92m wide, with semi-circular roof over 2.28m vertical sides-concrete lined Length: 2286m

VI. Power Generation
Mainpower house-underground: 540,000 kw

The Koyna dam is located in the peninsular shield of India, which was considered to be free from any seismic activity. The seismic zoning map prepared by the Bureau of Indian Standards projected it in the zone of very low activity. A horizontal design seismic coefficient of 0.05g was, therefore, adopted for design.

The non-overflow section of the dam is very wide at the top. It is about 15m wide compared to a normal width of about 7.5m. The downstream slope of the dam is 0.153:1 compared to 0.7 or 0.75:1. The sleep slope continues to a depth of nearly 35m before the profile hits a more conventional slope of 0.725 to 1.0. This peculiar section of the non-overflow part of the dam was chosen due to the following reasons.

On the basis of the projections of the anticipated demand for power, the original proposal was to construct a dam about 82m in height to impound a volume of 2770 m.m^3, out of which 1,920 m.m^3 could be diverted towards the west. As it was believed that there was no demand for the complete power immediately, the dam as planned to be constructed in two stages. The first stage dam was to be about 62m high. Considering the rapidly increasing demand for power and the requirements of irrigation downstream, the construction of a dam section with a lower height was started. When the dam had reached beyond mid height, the restriction for the western diversion was relaxed and the section was modified to suit the full height. The new section was achieved by adding extra weight on the upstream side with the help of an almost vertical wall with a wide roadway at the top. The section was designed for no tension for an earthquake acceleration of 0.05g (Fig. 4.33). This was a section typically designed to take care of static forces; however, a very poor one from the dynamic point of view. As the Deccan plateau was considered to be most stable seismically, no further notice of the poor dynamic characteristics of the dam section was taken till the earthquake hit it. The section actually looked like a huge wall (35 m high) setting on top of a conventional concrete dam, making it vulnerable at the junction of the dam and the wall.

Main Structures

The Dam

The Koyna dam is a gravity structure, 85.34m above the river bed and 103.02m above the deepest foundation. Its length is about 853.44m. The dam is constructed in monoliths, generally 15.2m in width. The contraction joints between the monoliths are provided with two copper water-seals, with a formed hole in between. The formed holes have not been filled but left to serve as drains and connected at the bottom to the drainage gallery.

As stated above, it was originally intended to build a thinner section of the dam for a lower storage Full Reservoir Level of El. 2135 and to raise the height later by adding concrete on the downstream. During construction, when the dam had actually been raised to some height in the first stage, it was decided to raise the height in a continuous operation to the ultimate storage level (Full Reservoir Level 2158.5). This was effected by thickening the section above Els 2060 and 2065 in the case of the non-overflow monoliths. In the case of the overflow section, which was somewhat thicker, this was effected by providing an upstream overhang, at the top. The spillway has a radial gates of 12.5 × 7.6m. The flood lift is 1.98m over the Full Reservoir Level of El. 2158.5. The energy dissipation is by a hydraulic jump on a sloping apron with an end sill.

The foundation treatment for the dam consisted of shallow grouting to consolidate the foundation rock and deep grouting to provide a grout curtain on the upstream, followed by a row of drainage holes. Consolidation grouting was restricted to about a third of the foundation width, with holes spaced 7.6m in a diamond pattern. Grout intake was negligible.

Deep grouting was done from the foundation gallery, after the dam had risen to a considerable height. The grout curtain generally extended to a depth corresponding to 75% of the head of water on the foundation. Grout pressures ranged from 7.03 to 21.09 kg/cm^2, depending on the grouting stage. Spacing of grout holes was generally 3.04m.

The dam was designed as a gravity section with the following criteria:

1. Density of rubble concrete : 2672 kg/m^3
2. Density of water : 1020 kg/m^3
3. Density of submerged silt : 1020 kg/m^3
4. Density of earthfill on downstream : 1620 kg/m^3
5. Uplift assumption:
   (a) 100% of reservoir level at the heel
   (b) 66–2/3% of reservoir level at the foundation gallery
   (c) Tailwater at toe

The uplift was assumed to act on 100% of the area.

The design requirement was that there shall be no tension under normal loading conditions. The section was checked for a horizontal acceleration force of 5%g with the reservoir fill level. A small tension was considered permissible in this condition. The sections were checked for the following loading conditions:

1. Reservoir level at Maximum Water Level of KRL 2165.0
2. Reservoir level at Full Reservoir level of KRL 2158.5
3. Reservoir level at Full Reservoir Level + Earthquake acceleration of 5% g
4. Reservoir level at minimum draw off i.e. KRL 2000
5. Reservoir level at KRL 2000 + Earthquake of 5% g.

The Intake

The intake system consists of the intake weir, the intake channel and the intake tower.

At the entrance to the intake channel, a masonry weir is provided with its top at KRL 1993. The length of the weir is 93.23m. The minimum draw-off level of the lake is KRL 2000.

The intake channel extends from the intake weir to the intake tower. It is located along the ridge between two streams, so as to prevent drawing in of trash or rolling stones during flash floods. The channel is formed by deep trench excavation into the rock; and a suitable gradient is provided from KRL 1980 at the weir to KRL 1960 at the intake tower (entry to head race tunnel). The intake channel is unlined, except in a short length of 100 ft (30.48), where brecciated rock was contacted; and in this portion, the sides were lined.

Geology of the Region

Western India is dominated by the Deccan volcanics, or traps, which crop out over an area of approximately 524,000 km^2 between Belgaum and Nimach, and between Saurashtra and Jabalpur. Although geological studies have been made of many areas, few systematic regional investigations have been undertaken of Maharashtra and Gujarat.

The volcanics consists mainly of basalts and are spread out over an irregular surface of Archaeab, Algonkian and Gondwana Rocks. They are likely to be at least 2,000m thick in the Lonavala area, but thin out towards the east and form only a thin veneer around Wardha and Nagpur.

The pronounced scarp of the continental divide, averaging 600 to 800m in height—except for the higher isolated summits which overlook the plateau—exposes magnificent sections of the Deccan lavas and Inter-trappeans.

Underneath the Deccan volcanics is the great peninsular shield which consists primarily of two main Archaean Provinces: the Dharwar Iron-Ore Province and the Charnockite-Khondalite Province.

Dykes

Cutting the lava-flows are numerous basaltic and doleritic dykes. These dykes are widespread in a dispersed form throughout the western plateau, but they are more concentrated in the Konkan near Mumbai. Here they have an inferred connection with the north-south Panvel flexure, in a manner comparable to the dyke swarm in south-east Greenland. The dykes are upto 90m in width and 30km in length.

Along the Narmada and Tapi Valleys, there is a dyke swarm 70km in width, which is oriented ENE-WSW, part of which may be related to the east-west folds of the Rajpipla lavas. One dyke is 56km in length, while some others reach a width of 60m.

In saurashtra, the dykes are partly radial to the plutonic centres, and partly in the form of a 50km wide ring between Girnar and Chanmardhi.

There are no swarms or dykes in the Koyna area, but sporadic dykes are found in certain regions. These dykes indicate a period of tension within the volcanics. Some may have originated with the lavas; others were evidently intruded at the end of the plutonic phase. They are cut by the fractures and the tensional stresses are thought to be connected with Cretaceous-Oligocene volcanic eruptions, whereas the fractures may have been of Miocene age, and subject to possible rejuvenation.

Folding

Over most of the Deccan, the volcanics are virtually horizontal, although dips of 1 (17m/km) are probably common. However, these dips are difficult to estabilish without the use of accurate levels, because even a hand Abney level is not normally satisfactory for determining dips of less than 1 . On the east-west spur which juts out from the plateau just south of Amboli Ghat (Belgaum-Vengurla road), there is an unmistakable eastward dip of the basalt of 1.50 to 2 . Important exceptions do occur when and where stronger folding is manifest. A few examples of such folding may be cited, but the list is not in any way exhaustive.

(a) In the area between Mumbai and Panvel, westerly dips of 5 – 15 are observed. In this area, there is a monoclinal fold, on a small scale resembling the monoclines on the southeast coast of Greenland and the east coast of Mozambique (Auden 1949). The northern exposed limit of the Konkan monoclinal fold is not known, but it may strike out into the sea between Dahanu and Daman. The evidences from the seismic reflection work undertaken in the Gulf of Cambay by the Oil and Natural Gas Commission indicate that the flexure is located under the sea. From an air inspection on the flight from Belgaum to Santa Cruz, a westerly dip of the lava flows is seen over a distance of about 80km between Srivardhan and near the northern end of the Alibag peninsula. It is presumed, therefore, that the monoclinal fold runs out into the ocean a short distance south of Bankot. The full extent of this fold, as exposed in the Konkan, may be of the order of 240 km.

(b) North and south dips of 40 are recorded along the Narmada.
(c) Dips are recorded in the area between Nandwa and Sausar.
(d) In Saurashtra, the dips are probably radial from the plutonic centres.
(e) In Kutch, the lavas are folded, together with the underlying Jurassic-Cretaceous and the overlying Eocene-Neogene. Here, the volcanics have spread into a region which became part of the Orogenic zone. South dips of upto 20 are known.

Faults

Major faults have only seldom been noticed in the main area of the volcanics, but an important exception is the east-west faulted zone, extending 70km, which has exposed the underlying basement around coordinates 20 25'N:78 E.

On land, no major faulting has been established in the Konkan at the foot of the great scarp. The scarp, as distinct from the coastline, is a feature with numerous embayments and promontories and totally lacks a straight geometric. Further, no fault has been observed along the north-south reach of the Koyna valley.

The line of thermal springs is at present the surest indication that a fault may exist in the Konkan, concealed below a cover of Deccan volcanics. These springs are confined to the outcrop of the Deccan volcanics, even though the waters presumably originate from faults, or eruptive channels, within the underlying Basement.

No thermal springs appear to exist south of the outcrop of Deccan volcanics. Even if a north-south fault exists in the Basement under the Deccan volcanics outcrop, the southward extension of such a fault has not yet been established in the Basement rocks which are exposed to the south. Detailed geological surveys have not been carried out in this region, and a fault system may eventually be located in the Dharwar outcrops, south of the Deccan volcanics.

It has been stated by some authorities that the Koyna earthquake was connected with a north-south fault, along which the Koyna valley north of the dam had been eroded. While north-south shear zone has been observed at the dam site where the valley is orientated NNE-SSW, this did not cause any vertical displacement of the lava flows. Moreover, there is no evidence from the air photographs of any major fault along the general trend of the river. If such a fault did exist, its extension to the south would have been expected and should be manifest in the hills 10km south of the dam, where the river turns eastwards towards Karad. No such fault is evident.

It is of interest that the epicentres of the main Koyna shocks of 13 September and 10 December GMT (11 IST), as well as of the aftershocks, fall upon a fairly well demarcated NNE-SSW line. This direction is divergent from the presumed NW-SE extensions of the Dharwar rocks underneath the volcanics, and the north-south to NNW-SSE trends along the coast. There is possibly some analogy here with the Nevada earthquake of 1954 in which the seismological evidence indicated a fault at depth striking west of north, whereas the surface indications were that the fault ran east of north. At Koyna, however, no surface fault has been observed.

On the basis of available information at that time, it was established that there was no discernible relationship between the Koyna earthquakes and the fractures and structures such were visible in the exposed upper crust.

Geology of the Dam Site

The axis of the dam is aligned 120 – 300 in a part of the valley which is oriented NNE-SSW. The dam is founded on Deccan traps, i.e. volcanic basalts, which in this area are virtually horizontal. Although the trap formations are almost horizontal, the actual interfaces between different volcanic types are irregular and do not remain at constant elevations

While the massive basalts have compressive strengths ranging from 1000–1500 kg/cm^2, the vesicular varieties containing zeolites and chalcedony have relatively lower strengths, lowest upto 100 and 200 kg/cm^2.

Except for 200m of the river section—between foot chainages 2575 and 3225—the higher parts of the dam are founded on the 33.5m flow of massive basalt. In the river section, it was, however, necessary to take the formation down through the 7.5–9m weak formation no. 2, and to base the dam on the lower massive basalt (formation no. 1) with foundation level, at about EL 1850 feet (563.9m).

A north-south fracture or shear zone cuts across the dam, at an angle of about 60 to the axis and crosses parts of Monoliths 12–13–14. This zone was also found under one of the piers of the bridge, just below the stilling basin, and under the retaining wall of the left bank. It is 4.5m wide where encountered in Monoliths 12, 13 and 14 at the top level within the 30.48m basalt layer, but tapered down to 1.8m width where it passed into the tuffs and vesicular traps at El. 1870. The rock in the shear zone was broken and slickensided and filled with gouge. There was no vertical displacement of the lavas on either side of the fracture. The zone was excavated down to El. 1880, backfilled with concrete and grouted. A gallery has been provided above the treated zone. An accelerograph is located in this gallery.

The dam site, tunnels, penstocks and powerhouse have been investigated by a very large number of borings. Nowhere has any significant fault been found, either along the Koyna valley, or westwards in the region of stage I to III of the Project.

Koyna Earthquake

After the impounding of water in the Koyna reservoir in 1962, small shocks were being recorded in the reservoir area; some of them also accompanied by sounds. In 1964, three seismic observatories were established around the reservoir at Mahabaleshwar, Satara, Govalcot and Koyna dam by the Central Water and Power Research Station, Khadakwasla (Pune).

The interpretation of seismograms was carried out for period from 10 September 1964 to 12 September 1967. The number of recorded shocks was 154, their magnitude from 2 upto 3 in Richter scale. Epicentres of these shocks were determined in the areas of the middle and Southern parts of the reservoir, mostly on its western side. Their foci were provisionally determined at the depth from 3 down to 5km, probably more.

Earthquakes of 13. September 1967

A large number of shocks occurred on 12 and 13 September within a small area around the Koyna dam. The two shocks of 13 September had magnitudes about 5–5.5 as determined from Indian seismic observatories. Their epicentres were established instrumentally at 17 24.0′ N and 73 44.8′E, near the Koyna dam. These shocks caused slight and moderate damage to buildings in Koyna Nagar and created small fissures in the soft soil. Outside Koynanagar, at a distance of few miles, even the mud houses in the villages were not affected. However, these shocks were felt within a radius of about 120 km in Patan, Satara and Karad in the east, Wai and Pune in the north, Khed and Chiplun in the west and Kolhapur and Ratnagiri in the south.

The focal depth of this shock was determined seismologically down to 4km. According to the grade of intensity and value of magnitude, the focal depth was larger, down to 10km or a little more.

From 13 September 1967 upto 11 December 1967, seismic observatories in the Koyna dam area recorded many small aftershocks of intensity upto 3 to 4 and magnitude upto 2 and 3. The major earthquake in this region occurred on 11 December 1967.

Earthquake of 11 December 1967

According to Indian seismic observatories, the shock occurred on 10 December, at 22–51–19.0 GMT. The epicentre was determined at 17 22′.4N, 73 44′.8E, i.e. about 3 km to the south from Koyna dam; depth of the focus was 8km. A perusal of the seismograms showed that there were two shocks in very quick succession. Koynanagar was almost razed to the ground, and the Koyna dam was damaged.

Magnitude of the Earthquake

The magnitude of the earthquake was determined by different institutions. From several determinations, it followed that the earthquake was of magnitude around 6.5m maybe a little more, upto 7.

Intensity and Intensity Distribution

The earthquake of 11 December was felt extensively in the Indian peninsula upto Surat and Ujjain in the north, Nagpur and Hyderabad in the east and Bangalore in the southeast. Radius of area of perceptibility was over 600km.

According to several computations, the mean depth of focus of the earthquake could be accepted from 10–12km down to 20km.

The earthquake of 11 December 1967 was caused by a sinistral strike-slip along a fault striking N26E, and dipping at an angle of 66 towards N296 E.

Seismo-Tectonic Considerations

The focal depth of foreshocks from 1964 upto 13 September was from 3 to 5km. The focus of the main earthquake of 13 September 1967 was at a depth of about 10km, and the hypocentre of the earthquake that occurred on 11 December 1967 was at the depth from 12 down to 20 km.

Such a special distribution of foci in a small area, testifies to the fact that from the depth approximately 3–5km down to the depth of 25–35km, there is an almost vertical fault zone. Epicentres belonging to the above mentioned foci of foreshocks and main shocks form a band on the surface, which strikes approximately from north to south. This indicates that the fault zone under the valley has the same approximately, north-south strike. Tectonic movements on it could cause the shocks in the Koyna dam area.

The movements of the supposed Koyna fault zone were not expressed in any disturbances of the superficial geological structures, as it often happens when the focal depth is large enough. The movements may have been lateral (strike-slip) and/or their vertical amplitude was possibly small.

The strike of the supposed Koyna fault zone, coincides approximately with the strike of:

(i) Areas of intensity distribution of 11 December Earthquake;
(ii) Western Ghats; and
(iii) The line of hot springs, which trends from the Mumbai area of the Chiplun region.

Induced Seismicity by Reservoir Loading

Ever since earthquakes (to magnitude 5) occurred in 1936 with the 80% filling of Lake Mead, engineers and geoscientists have been concerned about whether a deep impoundment of water can induce macroseismicity at a geologically susceptible site. Although there is no universally accepted proof that this can occur, after tens of site studies and analyses, strong circumstantial evidence does exist for it at several major reservoirs in the world. However, there is a growing consensus that reservoir-induced earthquakes are limited and probably should not be considered except for very large and deep reservoirs with select geological settings.

The milestone earthquake activity of Lake Mead and Hoover Dam region was analysed by Carder (1945, 1970). His principal findings correlated the release of seismic energy with the water load and movement of young faults at depth within the regional fault-blocks. The concern for induced earthquakes was further strengthened by events in the 1960s at or near several large reservoirs, such as: Kariba, Zimbabwe (M = 5.8); Kremasta, Greece (M = 6.3); and Koyna, India (about M = 6.5), although this last particular region has been considered to be aseismic. The apparent increase in seismic activity occurred subsequent to reservoir impoundment and it was inferred that loading caused or triggered the earthquakes. The growing number of possible reservoir-loading occurrences resulted in many separate investigations, which are reported by Judd (1974), Milne (1976) and Rogers and Le (1976).

Reservoir-induced seismicity is the occurrence of earthquakes that are triggered in one of several possible ways by the operation of a deep reservoir. Unfortunately, a reservoir-caused earthquake has no discrete signature distinctive from a naturally occurring seismic event, that poses as a major obstacle to the identification and verification of induced macroseismicity Meade (1982) has made a critical and state-or-the-art assessment of the evidence for reservoir-induced events as well as field/ laboratory methods for collection and interpretation of data.

According to Meade's findings, less than 1% of the world's reservoirs have been associated with macroseismicity (magnitude greater than 3) and only 14% of reservoirs deeper than 95 metres are being considered. The response made to vague threats of induced seismicity is conservatism in the seismic design of a dam, which normally is not justified. Subjective judgement are often repeated and taken as fact. The cost of such unnecessary conservatism is high and whether the threat is real or imagined, the cost is an important factor.

The induced seismicity reported by many investigators has been based on the following- there types of circumstantial evidence, in decreasing order of importance.

- Post-impoundment increase in seismicity, which is most convincing where it is corrected strongly with fluctuations of the water level (as at Nurek).
- Correction of water level and seismic activity.
- Display of a distinctive aftershock pattern, as expressed by an unusually high slope of the magnitude-frequency relationship

Most macroseismic events are generated by tectonic forces, whereas microseismicity can be generated by either tectonic or non-tectonic forces, including loadings such as by reservoirs.

Eight of the most widely reported cases of induced macroearthquakes in the world are summarised in Table 4.1. In only four instances are damaging earthquakes believed to have been associated with the reservoir impoundment.

Table 4.1 Selected Cases of Induced Macroearthquakes (After Meade)

Reservoir	*Location*	*Depth* (m)	*Volume,* X10^{-6} m^3	*Largest* Earthquake*
Hoover RI	United States	166	35,000	$K_L = 5.0$
Kariba	Zimbabwe	122	160,000	$M_b = 5.8$
Kremasta	Greece	120	4,750	$M_s = 6.3$
Koyna	India	103	2, 780	$M_s = 6.5$
Kurobe	Japan	186	149	$M_s = 4.9$
Manic 3 RI	Canada	98	10,423	$M_b = 4.3$
Hsinfengkiang RI	China	80	10,500	$M_s = 6.1$
Nurek RI	Soviet Union	215	11,000	$M_s = 4.5$

*M_L : a local magnitude scale used
M_s : magnitude measured from amplitude of surface waves
m_b : magnitude measured from amplitude of body waves
RI : evidence supports reservoir-induced macroseismicity

Calculations indicate that the stresses due to the load of the water in the reservoir are too small to be able to fracture any competent rock. The most likely hypothesis is that the rocks near the reservoir are already strained from tectonic forces (so that existing faults are almost ready to slip) and the reservoir adds a strain perturbation that triggers a slip and/or increased water pressure resulting from the impoundment lowers the fault strength so that slip occurs from the applied tectonic stress.

A simple argument can be given for induced earthquakes with $M_L \geq 5$. In this case, the minimum volume of the strained region would require a focal depth z greater than 1 km. A fracture at this depth occurs under lithostatic stresses with the greatest principal stress $p_3 = \rho g z$ acting vertically, and

$$p_1 = p_2 = \{\nu/(1-\nu)\}p_3$$

where ν is the Poisson's ratio. The shear strength of a vertical fracture with coefficient of friction μ is then $S = \mu g \rho z \nu/(1-\nu)$, with stronger inclined fractures. For normal values of the parameters and $\mu = 0.5$, the minimum shear strength for a dry vertical fracture is, thus, about 50 bars at 1 km depth. If water, density e_w, permeates the fault, its strength is reduced by $(e-e_w)/p$.

Therefore, the minimum shear strength for a wet vertical fracture at 1 km depth is about 30 bars. By contrast, a 150 metres-deep reservoir produces a maximum shear stress of about 5 bars in the rock, which is an order of magnitude too small to induce earthquakes directly. It follows that pre-existing tectonic stresses are needed to exceed 25 bars to ensure that faults are almost at the failure point.

Earthquake mechanisms in examples of reservoir induction have been calculated in a few cases. For the main shock at Koyna Dam in India, the evidence indicates strikeslip motion; at Hsinfengkian Dam in China, the principal shock was also strike-slip, with a stress drop of about 10 bars. At Kariba Dam in Africa and Kremasta Dam in Greece, the mechanism was dip-slip on normal faults.

In contrast, earthquakes had thrust mechanisms at the Nurek Reservoir in Tadjikistan, USSR. There, more than 1800 earthquakes ($1.4 < M < 4.6$) occurred during the first nine years of filling of the 300 metres-deep reservoir. This rate is four times the average for the region prior to filling. Increased seismicity has occurred in a series of swarms; the two most dense being accompanied by rapid increases in the water level during the first two stage of filling, to 105 metres in 1972 and to 205 metres in 1976. All periods of high seismicity occurred when the water level was higher than it had been previously or was within 10 metres of its previous maximum. All of the largest earthquakes followed decreases in the rate of filling of the reservoir. As the reservoir approached its maximum size (extending 40 km upstream from the dam), the area of induced seismicity also increased.

Evidence Conducive to Induced Seismicity

Circumstances evidence for induced seismicity alone does not provide the basis for an unqualified, positive, deterministic decision. Currently, the evaluation of evidence is based to a considerable extent on subjective judgment; but engineers need to take decisions based on facts. Consequently, adequate data and field-experienced, mature professionals are critical to decisions on reservoir-induced seismicity.

The dominant findings and parameters that support or limit the evidence for induced seismicity by reservoir loading are as follows (Meade, 1982 and Simpson et al. 1981):

- Induced seismicity appears to be a transient phenomenon; usually activity has decreased with time and none of the largest induced events has occurred more than 10 years after the reservoir was filled for the first time. After some tens of years, the region around a reservoir reaches a new equilibrium and exerts no more effect on the seismic regimen than a natural lake.
- A deep reservoir in a steep-walled, V-type canyon of a reasonably young age with associated fault zones is a prototype setting for induced seismic activity.
- Induced seismicity can occur when less than 50 metres of water is impounded, but the stronger seismic activity occurs when the water depth is over 100 metres.
- The weight of water alone on the crustal rocks is not a causative factor. The downward movement of reservoir water can increase hydrostatic pressures in the rock mass thereby, increasing the pore pressure along a pre-existing fault plane under critical stresses.
- The addition of water pressure to a critically stressed fault plane may decrease the shear-friction values of the fault sufficiently enough to trigger a seismic event.
- A reservoir-induced earthquake has no discrete/unique signature that is distinctive from a naturally occurring seismic event. Furthermore, changes and/or migration of seismicity over a period of time are to be expected. The accuracy of epicentres and hypocentres is critical for an adequate evaluation.
- Statistical techniques can be used to determine any significant change in seismicity coincident with impoundment; simple distribution-free tests can evaluate the change before and after. However, correlation evidence of induced seismicity cannot be properly evaluated using statistical techniques alone. For example,

the response of the seismic parametres may be different during rising and falling water levels, but the correlation statistic cannot account for this difference.

- Evidence for induced seismicity cannot be based on b-values alone, as they are time-varying and sensitive to data accuracy, magnitude bias and method of calculation. Furthermore, similar b-value trends occur in natural seismicity.
- There is no unequivocal case in which a damaging earthquake (M6) has been triggered by reservoir impoundment.

Effect of Earthquake in Koyna Region

The earthquake which took place in the early hours of the morning of 10 December 1967 claimed about 180 lives, injured several hundreds and rendered thousands homeless. An old stone masonry arch bridge at Helwak near Koynanagar collapsed completely and a few cracks in the road to Koynanagar were observed. Damage to parapet walls on the road and bridge and culverts was extensive. The most extensive damage, however, was to buildings. Apart from the dam, the most critical structures of the project such as the intake tower, the headrace tunnel, the surge system, the pressure shafts and the other tunnels did not show any visible damage. There was only slight cracking in the powerhouse, which was not considered to be very significant.

Damage to the Dam

The damage to the dam was indicated by the collapse or dislocations of some components and an observed increase in cracking and seepage at various locations. The description of the damage can be discussed under the following heads:

(a) Damage at the top of the dam
(b) Damage on the downstream and upstream face
(c) Damage to the vertical shafts and the gallery system and control chambers
(d) Damage to the transverse joints

(a) Top of the Dam

The most violent movements occurred at the top of the dam. Over the Monolith No. 18, which was partly overflow and partly non-overflow, there was an 18m-high hoist tower, which was elaborately designed, from the architectural point of view, making the top wider than the base—something like a ship. It was an R.C.C frame with precast paneling. Most of the precast blocks were completely dislodged from the structure and fell wide apart. Due to this, even the RCC main frame had cracked at several places. Similarly, a reinforced concrete wall of the control panel over Monolith No. 24 and temporary masonry structure over the Monolith No. 17 showed a degree of damage. The spillway bridge and the supporting piers showed considerable damage. There was a longitudinal crack on the deck slab. The bridge was a two T-beam RCC structure without any cross beams. Many tiles on the footpath had shifted. The bearing of the bridge had shown a displacement relative to the bridge. Some cracking was observed on the downstream top corner of the ladder was bent at the centre due to the earthquake.

On the basis of the observed movements, an acceleration of the order 0.63 was estimated the Koyna dam at the top. The tiltmetre and seismometre installed at the top overturned and records could not be obtained.

(b) Downstream and Upstream Faces

On the downstream face, there was a sudden change in slope from 0.163:1 to 0.725:1 at El 627.88. This was the location at which extensive cracking and seepage was observed on the downstream face. The spillway blocks did not show any cracking. There was a no-vertical crack either. Horizontal cracks at El. 627.88 in the non-overflow section were in blocks adjacent to the spillway, i.e. blocks 13 to 18 on the right side and 24 to 29 on the left side.

The upstream face was investigated by divers using a television camera. Cracking had been reported in Monoliths 10 to 18 roughly around elevation 627.88m, i.e. between El. 621.78m and El. 633.98m.

(c) Vertical Shaft, Galleries and Control Chambers

The vertical shaft or the hoist well, which provided access to the galleries, provided a clear evidence of the cracking inside the body of the dam, particularly in Block 18. At about El. 627.88m, there was a crack all around the shaft and water could be seen oozing from it. Connection between the upstream reservoir water and these cracks as well as cracks in Block 16 to 18 and 25 to 27 was confirmed by dye tests.

Fine cracks were noticed on both sides of the foundation gallery. Some cracks appeared at the upstream top corner of these galleries and water droplets were seen oozing out from some of these. The cracks in the inspection gallery were mainly dry and less prominent, although the pattern was similar.

In the control chamber for the hoist for sluice gate operation, which formed an enlargement of the operation gallery, a prominent crack appeared on the upstream wall at about mid-height, which brought some trickle of oozing water.

(d) Transverse Joints

Seepage through the contraction joints of Monoliths 18 and 19, 26 and 27, 16 and 17 had increased considerably after the earthquake. The total water pumped from the sump in the foundation gallery, which was 273 litres per minute, almost doubled after the earthquake. In addition, spalling of concrete has been observed along the vertical joints in many monoliths. The contraction joint between Monoliths 26 and 27 was observed to have narrowed at the top and widened at the bottom.

Instrumental Observations and Data

Instrumental data on ground acceleration and the response of the dam is very important, i.e. of prime importance in quantitative evaluation of the crack patterns and dynamic structural behaviour of the dam. For the Koyna Dam, some data was available on accelerations during the earthquake as well as data on instruments such as uplift, stress and deflections before and after the earthquake.

Ground Movement and Acceleration

To record the ground movement and the response, seven instruments were installed in the Koyna dam. These tools included two 3-component Accelerographs in Monoliths 1A and Monolith 13 over the shear zone. In Monolith 1A, in addition, were installed one Bemiof seismograph with photographic recording, another of the same type with pen recording and one Wood Anderson type of seismograph. Two seismic types of tiltmetres one each in Monolith 10 and Monolith 17 were also installed.

Due to the power failure and some other causes recording of only one instrument, i.e. the accelerograph in Monolith 1A only could be obtained. This accelerogram as finally accepted is shown in Fig. 4.34. Figure 4.34c shows the position of the block 1A and the location of the accelerograph. The records show that the peak horizontal acceleration in the transverse direction was 0.51g, horizontal longitudinal acceleration 0.66g and the vertical acceleration 0.36g. It should be noted that this is the response of monolith 1A and not directly of the ground acceleration.

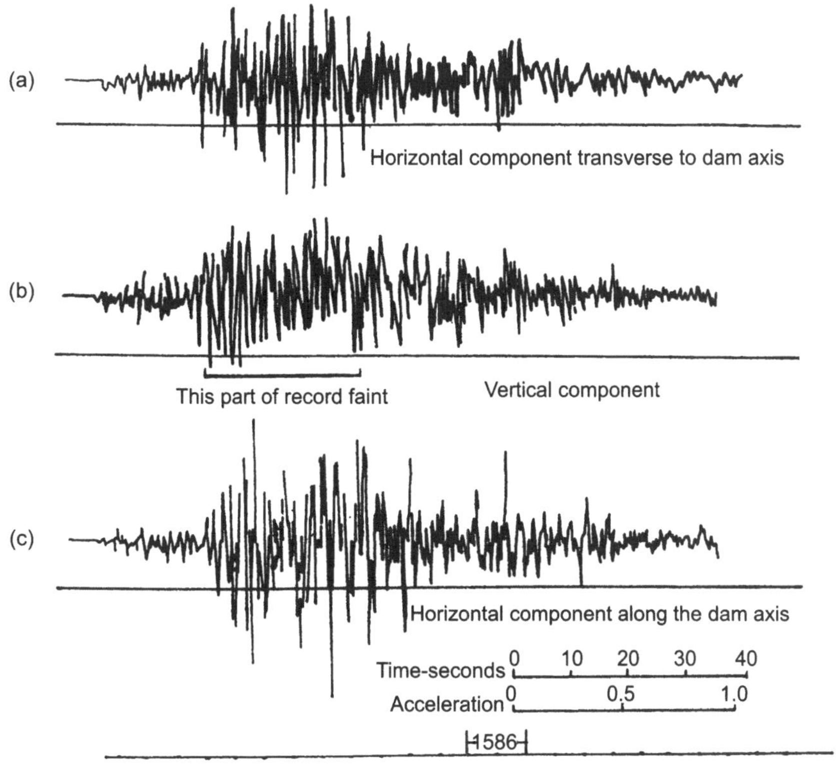

Fig. 4.34 Accelerogram recorded at block 1A of Koyna dam on 11 December 1967 at 04.21. IST

Static Behaviour from Instrumentation

As no continuous recording of the instruments such as uplift pressure cells, plumb bobs, strainmetres and stress metres were available, it was not possible to have any assessment of the dynamic behaviour of the dam from these instruments at the time of the earthquake. However, some of these instruments have indicated a residual effect of the intense shaking which persisted even after a week. A comparison of reading taken on 7 and 16 December (in between which the earthquake took place) for uplift pressures and the deflection readings are shown in Fig. 4.35 and Fig. 4.36.

The Koyna dam is located in the southern part of the Deccan Trap (Volcanic) region overlying a pre-Cambrian formation. The seismic zoning map of India, prepared by the 151 in 1962 and later revised in 1966, showed this region to be aseismic. After the occurrence of the Koyna earthquake, a large number of investigations on the seismicity of the

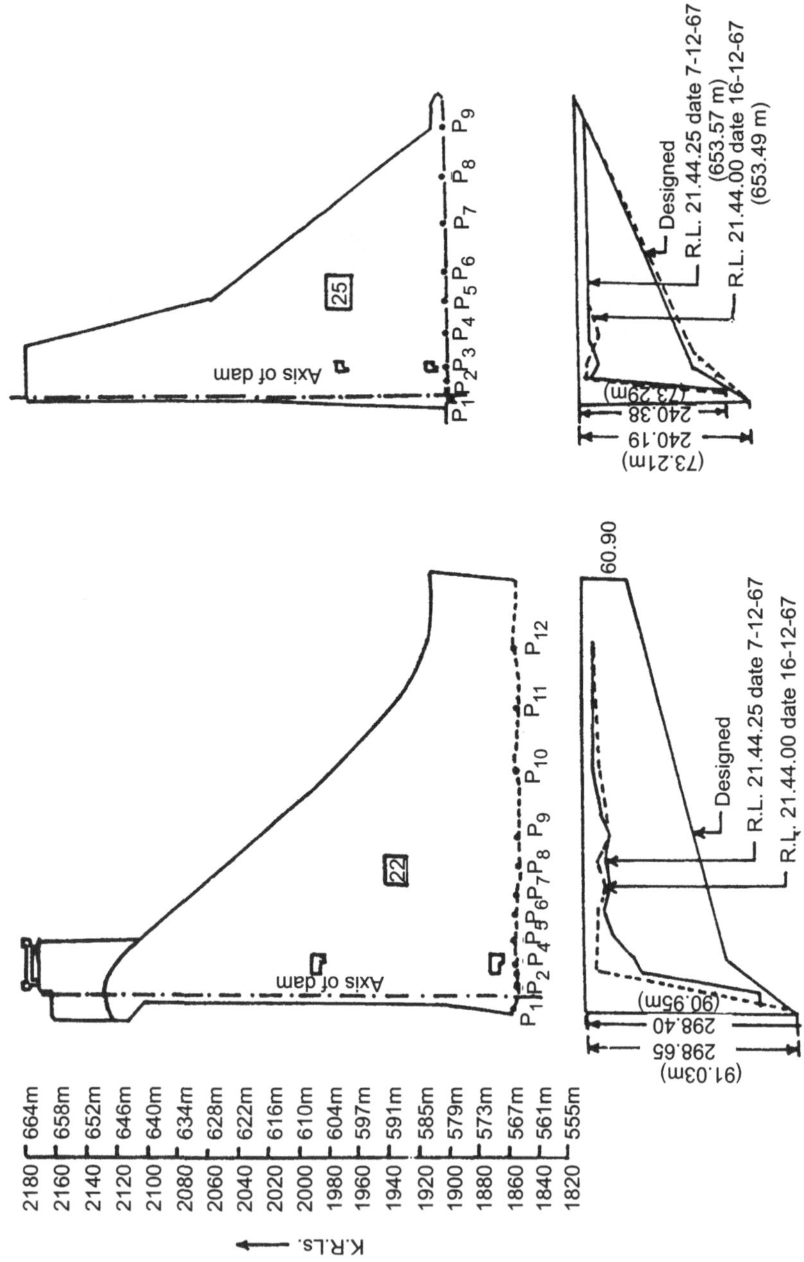

Fig. 4.35 Koyna dam—Uplift pressure diagram

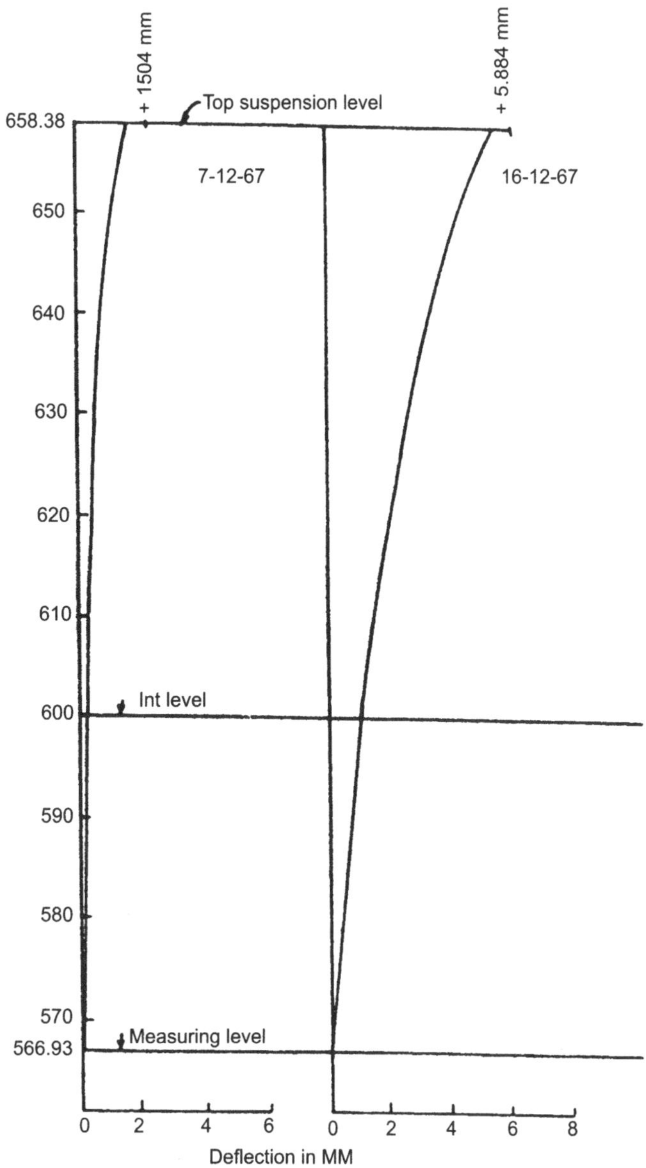

Fig. 4.36 Koyna dam—Deflection curve from monolith 22

region and a deep seismic survey have been carried out. The isoseismal map based on the damage due to December 1967 earthquake is shown in Fig. 4.37. Seismologies have now declared this region to be seismic and it is now located in the Seismic Zone IV with a basic seismic factor α_s = .05 and α_n = 0.10 as shown in Fig. 4.38.

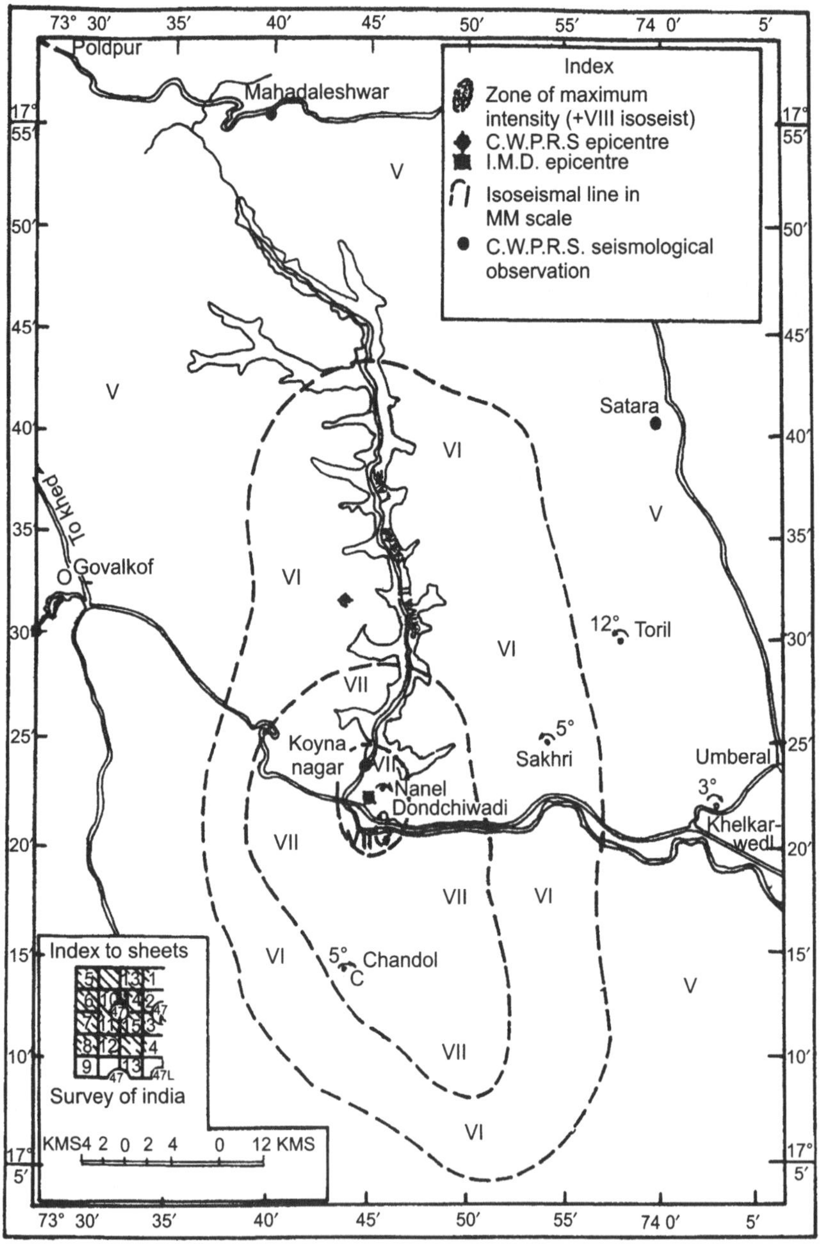

Fig. 4.37 Map of the Koyna region showing the two epicentres and isoseismals

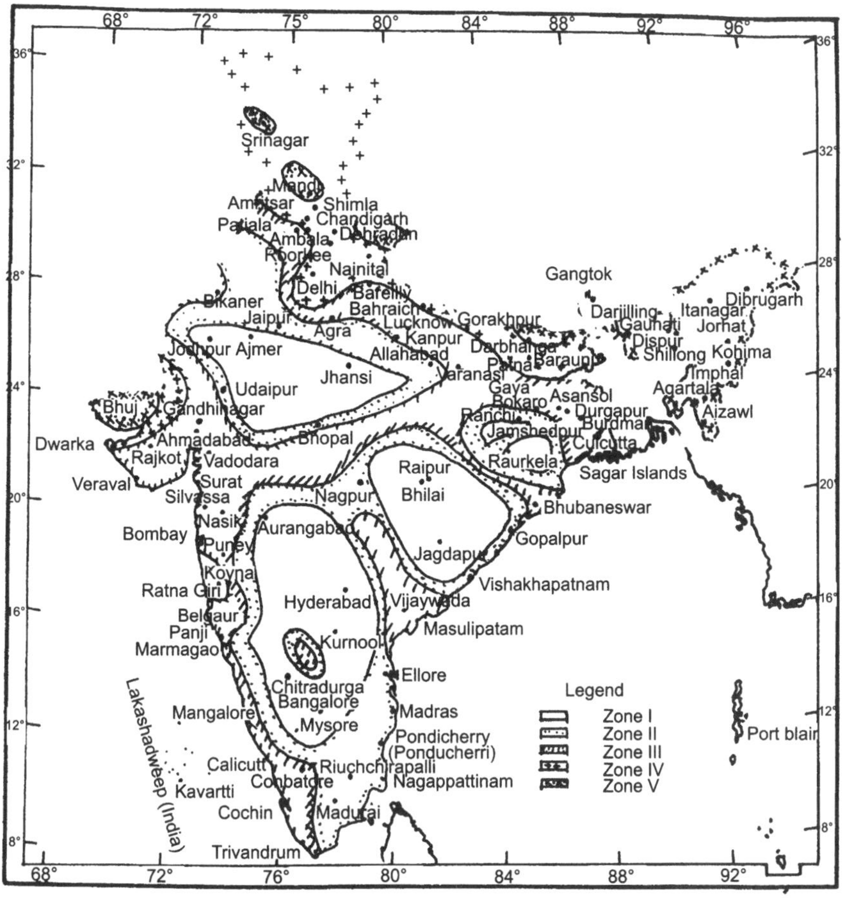

Fig. 4.38 Map of India showing major seismic zones

The magnitude of the December 1967 Koyna earthquake had been estimated in a range of 5.9 to 7.5. However, a value of 6.5 has been generally accepted. There is also some controversy about the accelerogram obtained from accelerograph located in Block 1 A of the dam which was an abutment block partly covered by fill. Some of the investigators have taken it to represent the ground acceleration, while others, as already stated, have made an attempt to back calculate the ground motion using the dynamic characteristics of this block. The record itself was rather indistinct and individual readings also differed to some extent. The duration of the strong shaking was 5 seconds and the accelerogram showed eighteen crossings per second. This latter value shows that the

Koyna accelerogram has higher frequency components compared to typical accelerograms recorded in California.

As regards the structural response of the non-overflow section of the Koyna dam to the earthquake, all studies, from the simplest using the stadolas principles to the most extensive ones using the finite element method and displacement time history computations, lead to the same and very significant conclusion. That is, the highest dynamic stresses arise at the downstream face where the batter changes from 0.725:1 to 0.1533:1.

A high tensile stress also occurs at the same level on the upstream face. Estimates of the tensile stresses at these upstream and downstream points are of the order of 40 kg/cm^2, respectively, by the Stadolas rapid and simple method. These values are valid only for the dynamic inertial load due to the mass of the dam alone. Chopra's detailed analysis, which includes the static as well as dynamic forces, gives the corresponding values of the order of 35 kg/cm^2. Hydrodynamic pressures have been estimated to increase these stresses by about 45%. Taking the ultimate tensile strength as 35 kg/cm^2, i.e. one-tenth the measured ultimate compressive strength, there is no doubt why the non-overflow monolith failed at this section. The profile adopted on static consideration was not appropriate and on dynamic condition not even qualitatively.

Due to the sudden increase of the shearing forces and tension on the upstream, the type of cracking in interior of the dam could be explained qualitatively. Quantitative analysis of the problem still remains difficult because of the serious problems in estimating the damping factor, particularly after partial cracking and the uncertainty in evaluating the dynamic tensile strength of the dam material which is necessarily constructed in lifts. Only future research in dynamic characteristics of dam material may throw more light on this issue.

Restoration Measures for the Koyna Dam

The Koyna dam is the key structure of the Koyna hydroelectric project, supplying power to an important industrial region in the country. It had an installed capacity of 540 MW at the time of the earthquake. To maintain a continuous supply of power, it was decided to strengthen it in two stages. In the temporary stage, prestressing and epoxy grouting of the cracked portion of the non-overflow monolith was resorted to and in the permanent stage, the blocks were backed with mass concrete. The basic considerations on which the decisions for restorative measures were taken by the designers are briefly reviewed in what follows.

Temporary Restorative Measures

Since any permanent measure to strengthen the dam were going to take at least a couple of years, it was necessary to resort to some temporary measures to ensure as much security as possible. It was necessary that the power supply be continued and, therefore, the reservoir level should be maintained well above the headrace tunnel. A detailed inspection of the dam showed that the spillway section, which had an over-hang, showed remarkably small distress. There was some wetness along the corners in the galleries at the foundation level and at the intermediate level and a fine crack at the valve chamber where the distance from the upstream was small. The most significant damage observed was in the non-overflow blocks. Most of these blocks showed a crack both at the upstream and at the downstream face near about the level of the kink. In many of the blocks, water could be seen seeping through these cracks.

For temporary restorative measures, therefore, the following decisions were taken on the basis of the above observations and other supporting data:

1. It was not necessary to strengthen the overflow sections which showed little distress.
2. In the non-overflow sections, the greatest distress was at the level of the kink where analysis showed high dynamic stresses and the shear friction factor was the minimum.

It was decided first to close the crack at the upstream end by epoxy grouting and, at the same time, pre-stress the top portion of the block by stressed cables as shown in Fig. 4.39. These cables served a double function of not only increasing the normal stress and thereby increasing the shear resistance but also enabled anchoring the top portion of block to the main body, thus, preventing large displacements, Initially, it was proposed to anchor the cables to the foundation. However, as the time available was short for construction, it was decided to anchor the cables at an intermediate level in the body of the dam. The further upstream could the cable be located, the more effective it was in stressing the upstream face but this was also likely to introduce tensile stresses below the anchor zone. After a number of tests, both on micro-concrete models and photoelastic models, the optimum distance of the meters from the upstream face was decided upon. Figure 4.39 shows the temporary remedial measures.

Permanent Restorative Measures

Pre-stressing by cables was a small measure and could not be a final solution to the problem. For a permanent solution, two alternatives were considered in the initial stage. The first one consisted of a backing of the

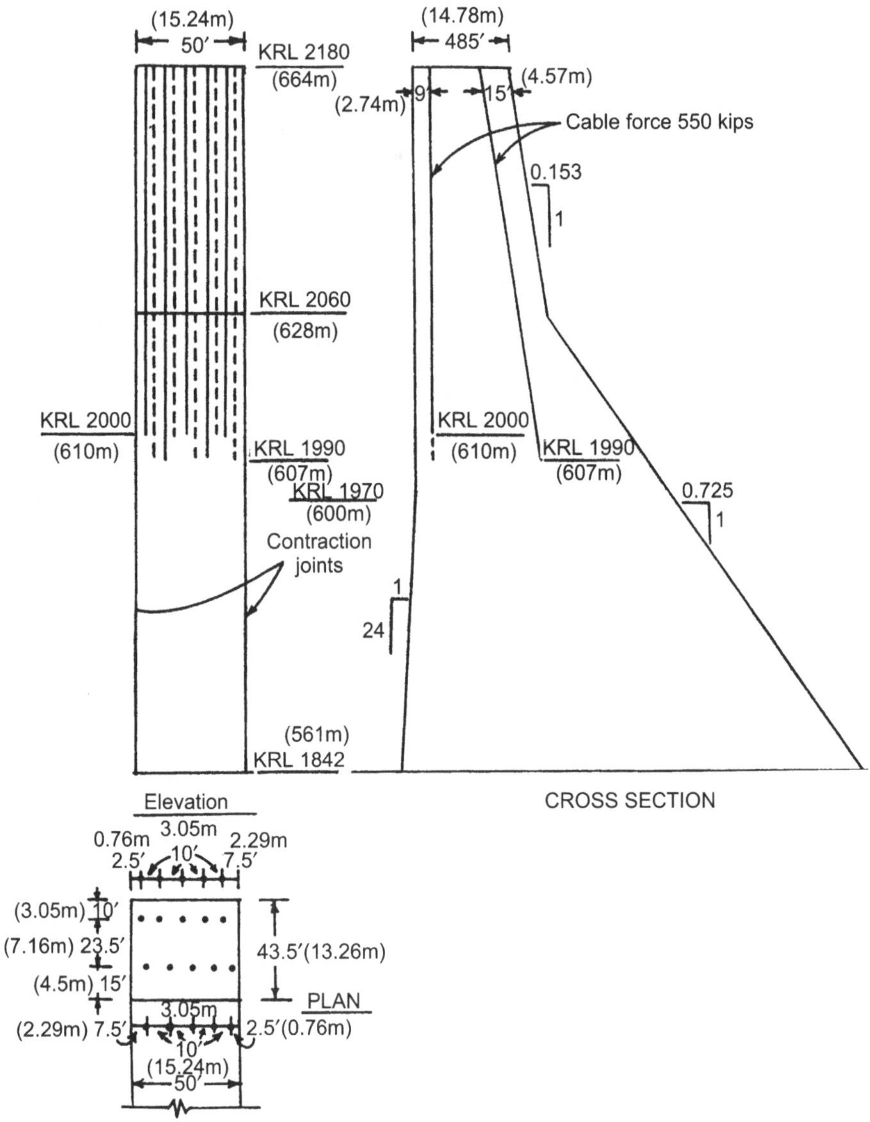

Fig. 4.39 Temporary remedial measures

downstream face of the dam by earth or rockfill. This strategy was rejected on the ground that the conditions at the interface of the rock and the concrete could not be easily visualised and the behaviour of such a composite dam under earthquake loads would be very difficult to estimate. The solution finally considered was to back the downstream face of the non-overflow section by concrete.

A number of studies, using mathematical and physical models of various types, were carried out. At the initial design stage, a rapid experimental method based on Stadola's static analogue was used to determine the mode shapes and frequencies and later, the response spectra method in conjunction with photoelastic method to obtain the stresses. The results are indicated in Fig. 4.40. The final design was based on a Japanese work which involved the determination of the frequency and mode shapes of various blocks by constructing gelatine models and vibrating these on a shaking table.

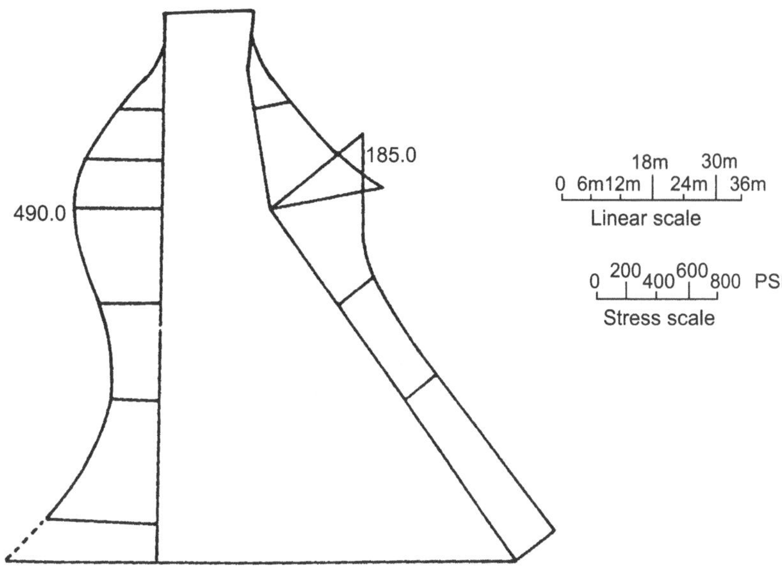

Fig. 4.40 Stresses at the boundary of the dam

The horizontal seismic force, which acts on the monolith at any elevation, was obtained by multiplying the response acceleration with equivalent mass. Bending moments and shear were calculated by the tradition methods. When the computed values were compared with the traditional method, it was found that the stress condition in the dam could approximately be reproduced by adoption of an uniformly distributed transverse coefficient of 0.5. Although the compressive stress during the earthquake was well within the ultimate strength of concrete used (175 to 225 kg/cm^2), the tensile stress exceeded 40 kg/cm^2 even after taking a damping factor as high as 12%. This was much more than the expected tensile strength. However, the dam had stood well except for the distress at the level of the kink. Consequently, it was decided to

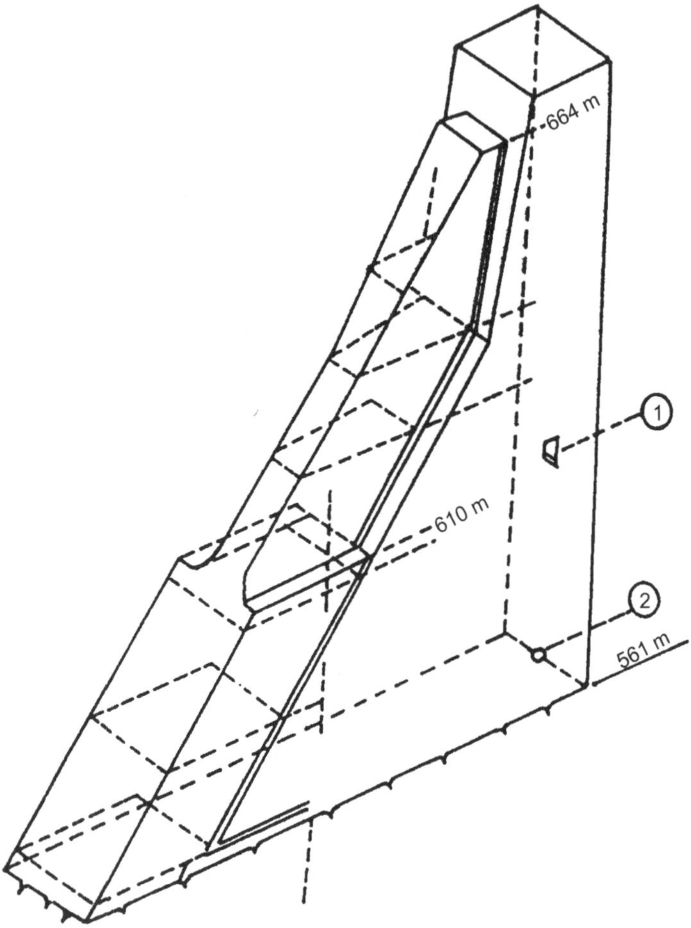

Fig. 4.41 Permanent remedial measures

adopt the conventional method of design adopting the non-uniform seismic coefficient of 0.15g – 0.20g for the design of the restorative measures. This decision was based on the observations on various dams in Japan as well as an assessment of the nature of damage that had occurred in the Koyna Dam. Redesigned section of the dam is shown in Fig. 4.42. To same time and ease in construction, it was decided to limit the backing on the entire face upto the level indicated, above which backing was in the form of a buttress.

The design ultimately involved the following assumptions:

(i) The Koyna dam is not likely to be subjected to any earthquake greater than the one that had occurred on 11 December 1967.

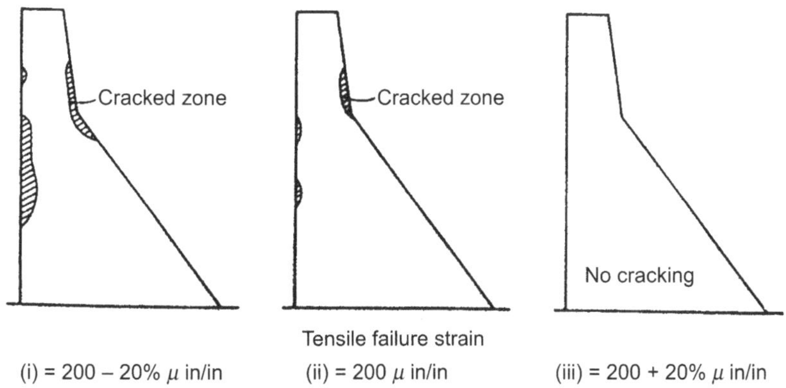

Fig. 4.42 Cracked zones in the dam for different tensile failure

(ii) The observed behaviour of the non-overflow monoliths was entirely satisfactory and, therefore, no strengthening was considered necessary.
(iii) At analytical and experimental work as well as observation showed that the kink in the dam was not desirable. The downstream slope of the dam should be straight as far as possible with the batter of the order of 0.8:1.0.
(iv) The standard codes of practice with non-uniform seismic factor ranging from 0.15g to 0.20g could be used to design restorative measures.
(v) Modifications could be made in the downstream backing to enable speed and ease in construction and that the effect of longitudinal vibration of buttresses was negligible.

Analysis and Design

The analysis and design of dam for earthquakes has progressed from empirical rule-of-thumb methods, through static force methods involving the use of seismic coefficients, to procedures that now recognise the dynamic nature of the problem.

Realistic analysis of the seismic response of dams were not possible until the development of the finite-element method, recent advances in dynamic analysis procedures, and the availability of large-capacity high-speed computers. Initially, all non-linear effects, including those associated with construction joint opening, concrete cracking and water cavitation, were ignored and the interaction effects of the impounded water and foundation rock were either neglected or grossly simplified. Subsequently, special techniques were developed to incorporate the interaction effects in linear analyses. The refined analysis procedures have been implemented to varying degrees for the different types of dams.

Shortcomings in Traditional Analysis and Design Procedures

Although new design criteria are now available, traditionally, concrete gravity dams have been designed and analysed by very simple procedures. The earthquake forces were treated simply as static forces and combined with the hydrostatic pressure and gravity loads. The analysis was mainly concerned with overturning and sliding stability of the monolith treated as a rigid body and with stresses in the monolith calculated by elementary beam theory.

In representing the effects of horizontal ground motion transverse to the axis of the dam by static lateral forces, neither the dynamic response characteristics of the dam-water foundation rock systems nor the characteristics of earthquake-ground motion were recognized. Two types of static lateral forces were included. Forces associated with the weight of the dam were expressed as the product of a seismic coefficient which was typically constant over the height, with a value between 2 to 0.10 and the weight of the portion was being considered. Water pressures, in addition to the hydrostatic pressure, were specified in terms of the seismic coefficient and a pressure coefficient based on assumptions of a rigid dam and incompressible water. Finally, interaction between the dam and the foundation rock was not considered in computing the afore-mentioned earthquake forces.

The traditional design criteria require that an ample factor of safety be provided against overturning, sliding and overstressing under all loading conditions-compressive stresses should be less than the allowable values. Often, tension was not permitted and even if it was, the possibility of cracking of concrete was not seriously considered. It had generally been believed that stresses were not a controlling factor in the design of concrete gravity dams; so the traditional design procedures were concerned most with satisfying the criteria for overturning and sliding stability.

Assuming the linear structural behaviour, the dynamic response of the tallest non-overflow monolith of Koyna Dam to the Koyna ground motion was analysed. The response results indicated larger tensile stresses in the upper part of the dam, especially around the elevation at which the slope of the downstream face changes abruptly. These stresses, which exceeded 40 kg/cm^2 on the upstream face and 60 kg/cm^2 on the downstream face, were approximately two to three times the tensile strength of the concrete used in the upper parts of the dam. Hence, based on the analytical results and strength data, significant cracking could be expected at locations consistent with the locations of damage caused by Koyna earthquake. The maximum compressive stress in the monolith exceeded (not shown in Fig. 4.40), which is well within the compressive strength of concrete. A similar analysis of the overflow monoliths predicted that the earthquake should have caused little or no cracking in these monoliths, which is also consistent with the actual damage. Figure 4.45 shows the cracked zones based on different tensile failure criterion.

Pseudo-dynamic Analysis of the Dam and its Buttress

The analysis of the dam and the buttress was carried out in the following steps:

1. The vertical stresses at the foundation level for the deepest section were computed due to the:
   (i) Dam only (gravity);
   (ii) Water pressure only; and
   (iii) Combined dam and water pressure along with uplift.
2. It was been whether the stresses are within limits, and whether the factors of safety against overturning and sliding are also within limits.
3. The effect of earthquake was then taken into consideration. A linearly varying coefficient (trapezoidal) was considered. The forces and moments at the particular elevation of analysis were calculated by simple integrations.
4. Thereafter, the water level was assumed to be at a low level and the buttress on the downstream side constructed. In actual practice, the buttress was constructed away from the dam and the gap between the two grouted later on.
5. Shear keys and dowels were designed to ensure proper transfer of shear. Elaborate analysis was done to calculate the normal and shear stresses along the contact surface of dam and buttress.
6. The water level was then raised and the stresses calculated again, due to:
   (i) dam and buttress (gravity);
   (ii) water pressure and uplift; and
   (iii) earthquake forces

It was ensured that the dam and buttress remained safe in the reservoir full with earthquake conditions.

REFERENCES

Bureau of Reclamation (1974). Earth Manual. U.S. Department of the Interior, 810 pp.

Carder, D.S. (1970) 'Reservoir Loading and Local Earthquakes' in Engineering Geology, Case Histories, Adams, W. (ed.) *Geological Society of America,* No. 8 pp. 51-61.

Casagrande, Arthur; Stanley D. Wilson, and E. Douglas Schwantes Jr, (1972), Baldwin Hills Reservoir Failure In Retrospect' *Speciality Conference,* SM&F Division, ASCE, April, pp 551-588.

Govind Rao, N.S. (1967): Failure of the Khadakwasla and Panshet Dams, *Journal of the Institution of Engineers* (India), vol. 47, N. 11, July pp. 1123-1144.

Govt. of Maharashtra (1961): *Report of the Commission of Inquiry for Enquiring Into the Failure of the Panshet and Khadakwasla Dams.*

Jansen, Robert B., Gordon W. Dukleth, Bernard B. Gordon, Laurence B. James and Clyde E. Shields (1967) 'Earth Movement At Baldwin Hills Reservoir,' *Journal of the SM&F Division,* ASCE, Vol. 93 No. SM4, July, pp. 551-575.

Judd, W.R. (ed) (1974) 'Seismic Effects of Reservoir Loading'. *Engineering Geology,* Amsterdam. Vol. 8, No. 1-2.

Kane, H. (1973). Confined compression of loess. *Proc. 8th Coni SM&FE,* Moscow, vol. 2.2., pp. 115-122.

Leonards, G.A. and Davidson, L.W. (1984). Reconstruction of failure initiating mechanism for Teton Dam. *Proc. Intn. Conf. Case Histories in Geotechnical Engineering,* St. Louis, Mo., May 7-11, Vol. 3, pp. 1103-1113.

Leps, Thomas M., (1972), 'Analysis of Failure of Baldwin Hills Reservoir' *Speciality Conference,* SM&F Division, ASCE, April, pp. 507-550.

Maswoswe, J. (1985). Stress Paths for a Compacted Soil During Collapse Due to Wetting. PhD Thesis, Imperial College, University of London, 485 pp.

Meade, R.B. (1982) 'The Evidence for Reservoir-induced Macro earthquakes' in 'State-of-the-Art for Assessing Earthquake Hazards in United States' Report No. 19, U.S. Army Corps of Engineers Waterways Experiment Station. *Misc. papers* S- 73-1 188 pp.

Milne, W.G. (ed) (1976) 'Induced Seismisity'. *Engineering Geology.* Amsterdam. vol. 10, no. 1-2.

Murti, N.G.K. (1967): Khadakwasla, The Oldest Masonry Dam, *ICOLD Istamboul,* Q. 34, R. 53, pp. 895-915.

Murti, N.G.K. (1977): Panshet Dam Failure—An Interpretation, *Journal of the Institution of Engineers* (India), vol. 58, July, pp 26-33.

Osterberg, Jorj O., (1989) 'Necessary Redundancy In Geotechnical Engineering', *Journal of Geotechnical Engineering,* ASCE, vol. 115, No. 11, November.

Penman A.D.M., (1953). Shear characteristics of a saturated silt, measured in triaxial compression. *Géotechnique,* vol. 3, No. 8, pp 312-328.

Proctor, R.R. (1933). Fundamental principles of soil compaction; also Description of field and laboratory methods. *Eng. News Rec.,* vol. 111, pp 245–248, 286–289 and 348–351.

Rogers, A.M. and Lee, W.H.K., 'Seismic Study of Earthquakes in Lake Mead, Nevada-Arizona Region', *Bulletin Seismological Society, America* vol. 66, no. 5, pp. 1657-1681.

Scott, R. F. (1987) 'Failure', *Geotechnique,* vol. XXXVII, No. 4, December.
Sherard, J.L., (1973). Embankment dam cracking. In: *Embankment Dam Engineering—Casagrande* vol. Wiley, New York, N.Y., pp. 271–353.
Sherard, J.L., (1985). Hydraulic fracturing in embankment dams. *Proc. ASCE Symp. Seepage and Leakage from Dams and Impoundments,* Denver, May 1985, pp. 115–141.
Sherard, J.L., (1987). Lessons from the Teton dam failure. *Dam Failures* (ed.) G.A. Leonards. Elsevier Press, pp. 239–256.
Simpson, D.W. and Negmatullaer, S.K. (1981) 'Induced Seismisity at Nurek Reservoir' Tajikistan, USSR. *Bulletin Seismological Society, America,* vol. 71, no. 5, pp. 1561-1585.
Walker F.C. and Daehn W.W. (1948). Ten years of pore pressure measurement. *Proc.* 2nd *Intn. Conf. SM&FE,* Rotterdam, vol. 3, pp. 245.
Water Power and Dam Construction (1985) 'Four Major Dam Failures Re-examined' *International Workshop on Dam Failures,* Purdue University, USA, 6–8 Aug.

Chapter 5

Investigations of Distress and Failures of Dams

INTRODUCTION

Preservation of water for human consumption—in a pot and then in pond—grew with necessity, just like any other commodity of daily requirement. Formation of ponds by digging pits at selected sites to conserve water for longer periods had their own limitations as these were mostly connected with the stream flows or natural springs. The more serious problem was withdrawing the water from these ponds as and when required. This problem gave rise to the formation of small levees over relatively flat lands, and limiting the height without endangering its stability. This strategy facilitated drawing water as and when required and in desired quantities. The height of the banks—sometimes on four sides—depended upon the material used and the experience with stability of slopes based on trial and error methods. Experience remained the only design in put and success was based on failures. Experience played a major role in the development of thumb rules and empirical approaches. With all round development of scientific knowledge, particularly the mechanics of materials, theories were generated from practices that made dam engineering a dynamic process and more responsive to environment. Failures got quantified by instrumentation of dams, which complemented observational practices and provided a firm basis of development in the design of dams. Design of dams progressed very fast, but understanding the causes of failures remained beyond comprehension so as to establish sequences and consequences in the right perspective. Even shortlisting the causes of failure and agreement over it by different investigators, perhaps, is getting more difficult with increase in knowledge and specialization bringing in more factors contributing to failure, than earlier known or identified.

DAM SAFETY GUIDELINES

The catastrophic failure of dams brought with it dam safety consciousness. The number of dams is increasing due to an increase in the demand for water for human consumption, agriculture, industrial development and power production. This is making the problems associated with design and construction of dams more compounded, as the location of dams at unfavourable sites has become unavoidable. This has added to the already existing hazardous situation, with respect to all around fast development of dams and problematic geological foundations, making it imperative to raise safety standards.

In Britain, the Reservoir (Safety Provisions) Act 1930 was passed by the Parliament after the failure of Dale Dyke earthfill dam on the evening of 11 March, 1864, during the filling of the reservoir which killed people around Sheffield. This Act, followed by the Reservoirs Act (1975), gave an impetus to instrumentation of British dams. In the USA, according to Jansen, a flood overtopped a dumped rockfill Lower Otay Dam in 1916. The downstream zone of about 40 m high dam failed and the central steel diaphragm, an unusual and unacceptable feature, was torn away. In the 1920s, there was considerable increase in dam construction activity. But there were also doubts about the safety of embankment dams, though these were found be economical compared to masonry and concrete structures.

There was difficulty in estimating maximum floods and due to short hydrological records American engineers thought of restricting the height of embankment dams. In 1925, 7.5 m high-Sheffield Dam in California failed under seismic forces, causing liquefaction of the foundation alluvium and the lower part of the dam. These events required closer monitoring and instrumentation of all the dams was considered, and experiments were initiated with Stevenson Creek Dam in California. As a result of the Baldwin Hills Reservoir in 1963, the California law for Dam Safety was strengthened, extending state supervision to offshore reservoir. The US Bureau of Reclamation increased its Dam Safety Activity after the failure of Tenton Dam in 1976. Since 1979, the US Army Corps of Engineers started compulsory surveillance of reservoirs during their very first filling. The laws of dam safety of California served as the basis for other states of the USA.

In India, dams and reservoirs are owned, constructed, operated and maintained by States, except for some organisations like Bhakara Beas Management Board, Damodar Valley Corporation, National Hydro-Electricity Power Corporation, who also *own* dams. A Dam safety Organization was formed in 1979 at the Central Water Commission of India, with the intention of assisting the State Governments in carrying out Dam Safety activity. Twelve States subsequently formed their own

Dam Safety Cells/Organisations. This helped to implement unified procedures and practices in dam safety to a great extent. In 1987, the National Committee on Dam Safety was formed by the Government of India to oversee dam safety activities in states and to serve as forum for exchange of views and to bring dam safety practices in line with ICOLD directives, consistent with India conditions.

As per the existing Guidelines for Safety Inspection of dams issued by Central Water Commission, India in June, 1987, dams have been classified in accordance with size and hazard potential as per Tables 5.1 and 5.2, respectively, in order to formulate a basis for selecting dams to be included in the inspection programme and their priority. Size classification may be determined by either storage or height, whichever gives the larger size category.

Table 5.1 Size Classification

Category	*Storage* (Hectare Metre)*	*Height** (Metre)*
Small	< 125 and > = 6	< 12 and > = 8
Intermediate	> = 125 and < 6250	> = 12 and < 30
Large	> = 6250	> = 30

*: The maximum storage elevation to be considered equal to the top of dam elevation.
**: Height with respect to the maximum storage potential from the natural bed of the stream at the downstream toe of the barrier.

Table 5.2 Hazard Potential Classification

Category	*Loss of Life (Extent of Development)*	*Economic Loss (Extent of Development)*
Low	None expected (non-permanent structures for human habitation)	Minimal (undeveloped to occasional structures or agriculture)
Significant	Few (no urban developments and no more than a small number of habitable structures)	Appreciable (notable agriculture, industry or structures)
High	More than few	Excessive (extensive community, industry or agriculture)

The guidelines provide for two phases of investigations since the scope and completeness of each investigation depend upon the availability and suitability of engineering data, the validity of original design assumptions and the physical condition of the dam. Phase I would comprise an inspection to assess the general condition of the dam and determine the need for any additional engineering investigations and analyses. It would consist of a visual examination of the dam and a review of the available engineering data, including operational records. It is not intended that costly explorations would be performed during Phase I inspection.

Phase II investigations are to be performed where the results of Phase I inspection indicate the need for additional investigations and studies. Phase II would include, as required, all additional visual examinations, measurements, foundation exploration and testing, material testing, hydraulic and hydrologic analyses and structural stability analyses deemed essential to evaluate the safety of the dam.

The guidelines merely present the procedure for investigating and evaluating existing conditions. Their scope does not encompass the studies and engineering solutions required for project modification to correct deficiencies found by the investigation.

Safety problem has also gained importance due to failure *of old dams,* which are mostly un-instrumented, ICOLD (1994) has defined ageing as deterioration due to 'time-related changes in the properties of the materials' used in the construction of dam and its foundations. These may be as a result of environment changes, including change in reservoir operation and lack of maintenance and periodic surveillance. Ageing affects the performance and functional objectives of dam, many a times beyond improvement, thus resulting in catastrophic failures.

INSPECTION AFTER UNUSUAL EVENTS

After an unusual event, there should be thorough inspection of the dam and its appurtenant works for detecting damage weakening, if any. After any earthquake occurs the dam shall be inspected as per ICOLD guidelines indicated in Bulletin No. 62 Inspection of Dams After Earthquakes, 1988. The technical inspection investigations should be conducted under the direction of competent professional engineers who are experienced in investigation, design, construction and operation of dams, applying the disciplines of hydrology, hydraulics, geomechanics, structural engineering and engineering geology. All field inspections should be conducted by qualified and competent engineers, engineering geologists and other specialists. All inspection observations shall be compared with the design assumptions and prediction, previous results, results of model studies and tests and limiting values, where specified. Any conditions that might adversely affect the safety of structures shall be closely observed. These conditions shall include excessive settlement, deflection, seepage, uplift pressure, deterioration of masonry/concrete, etc. Inspection of reservoir area, including reservoir rim, shall also be undertaken.

Structures, features, events and evidence that need to be examined during inspection are listed in Appendix.

MODERN TECHNOLOGY OF DAM SAFETY SURVEILLANCE AND EVALUATION

An advanced methodology has been introduced in Italy for many years now, to overcome drawbacks in the traditional checking methodology. This new methodology develops and makes more effective the checking criteria prescribed by current regulations/standards. The fundamental items of this advanced methodology can be synthesized as follows:

(a) The collection of measurements by means of automatic data acquisition systems, the most important requirements of which are flexibility, expandability, reliability and simplicity of maintenance. In addition to the acquisition systems of the static behaviour, automatic seismic acquisition systems are also often installed.
(b) The organisation of the measurements in computerised databases and their management and processing using specific computer programmes.
(c) The interpretation of the dam behaviour through mathematical models of the measurements collected both under normal operating conditions or in exceptional situations (earthquakes, floods, etc.).
(d) The real time checking of the structural dam behaviour ('On-line monitoring').
(e) The periodic check of some significant parameters of the dam and of its foundation, including in situ and laboratory investigations, so as to identify possible deterioration as well as to put into evidence possible inherent weaknesses dating back to the time of construction.
(f) A complete 'check-up' ('Certified Check') carried out periodically by specialists to evaluate the actual state of health of the structure and for the definition of the procedures to follow for an effective surveillance.

DAM SAFETY AND RISK ANALYSIS

Investment on dam safety can be justified to the extent of possible loss of property in case of dam failure but loss of lives cannot be quantified. Any compensation for loss of lives and property varies with cultural and social standards of a country and, hence, the cost input in dam safety varies from country to country. Similarly, risk analyses, though adjudged by well-tested design and construction practices adopted, helps to assess the probability of failures of dams and their adverse consequences. Serious objection to 'risk analyses' is that it precludes the use of engineering judgment and intuition, whereas many favour it, considering risk analyses and engineering judgment to be complementary to each other. Any analysis requires the quantification of several factors, from confidence in initial investigations to interpretation and extrapolation of field and laboratory data adopted for design. Commonly, risk analysis for dam safety design covers the following facets:

- the factors that can lead to dam failure are identified and the possibility of their occurrence and sequences are estimated.
- the adverse events that may initiate failure and potential modes of failure may be identified.
- the set of circumstances under which a dam may fail is evaluated.
- for each potential failure mode, the consequences of failure are identified that comprises sum of the estimated economic, social and environment losses.

In all these issues, the 'human error' which is one of the prime factors attributed to failure of dam and may occur at any stage, is not quantifiable, and so comprises the 'emotional trauma' that a person undergoes on various losses. Moreover, damages when rates as 'extensive' , 'significant' or 'minor' have no standard definition. The terms mean differently, depending upon the compensation capacity of the owner or the party concerned.

This dichotomy has led to 'legislation on safety of dams' and several countries usually have some form of legislation. ICOLD formed the Committee on Dam Safety in 1982 and developed dam safety guidelines (1987). *Personnel:* Modern dam engineering, surveillance and instrumentation of dams have raised the safety standards of dams provided efficient processing of data is followed. But skill and competence wedded with experience of personnel involved in surveillance cannot be substituted. Hence, the personnel involved in investigations and construction of dams and their surveillance get trained and develop 'observational sight' so as to reduce the probability of sudden failure and failure of dams without signs of distress. This group of personnel, with the background of this training and experience, are better suited for deployment during investigations of failure. This is a rare situation in any project and continuity of engineers, skilled and semiskilled persons in any capacity, is not a practice. Dam Safety Panels for a region or dam, to advise any safety measures based on surveillance reports prepared periodically, have replaced the earlier pattern.

DAM FAILURES

In general the failure of a dam is a complex phenomenon as there is not one cause responsible for the failure and to identify causes of failure and to prove their validity is not possible. Secondly, it is difficult to identify the factor that has caused initial instability and which many have led to a chain reaction, adding to the deterioration at every step, resulting in consequential failure of the dam. During design and construction, an army of engineers is involved, with varying experiences and expertise and time is not limited. Cases of 'time over run' or 'cost over run' in dam construction are not uncommon, inspite of the fact that construction commences with PERT or CPM charts on hand. In addition to this, the 'value engineering' concept has been introduced. But for investigation of failure, relatively very small group of engineers are involved, with varying experiences of dam engineering and dam failure. The eagerness to rehabilitate the dams as early as possible becomes the primary aim of the owner. Either for valid reasons, such as requirement of water for standing crop or running industry or for legal compulsions so as to cover up the facts of failure, investigations if any, aim to rebuild the dam as early as possible. Academic interest is overlooked and all probably causes are not studied. Financial constraints also do not allow complete investigations and the real reasons of failure are not revealed. As a result, in similar cases of failure, challenges posed to the investigator get new interpretation due to new compulsions which may not be technical as well. Leonards (1982) catergorically says that the failures have neither lead to more rational design methods nor methods of investigating failures have improved.

REASONS FOR DISTRESS IN DAMS

Distress in any structure, for whatever reasons, vitally affects the useful service life, performance and safety of the structure. This factor assumes an added importance in case of large structures, such as dams, wherein a structural failure due to distress may result in hugh losses of life and property as well as colossal structural rehabilitation costs thereafter. In view of this aspect and the fact that a majority of dams constructed worldwide require rehabilitation and repairs to varying extents, it is pertinent to adopt proper diagnostic and rehabilitation measures for countering the distress-related damaging effects on these structures, for enhancing their useful life, and more importantly, for reducing/ preventing losses connected therewith.

CONCRETE DAMS

The reasons for distress in a concrete dam may be one or more of the following and the distress may manifest itself in the form of cracks, seepage, leakage, pit formation, surface abrasion, rebar corrosion, etc.:

- Design/construction deficiencies
- Erosive forces
- Alkali-aggregate reaction
- Nature's fury—earthquakes/heavy floods/volcanic activity
- Exposure of dam concrete to aggressive waters
- Seepage/leakage
- Rebar corrosion
- Freezing and thawing
- Ageing of concrete
- Waterstop failure

Design/Construction Deficiencies

Incomplete or incorrect design data or criteria can result in an improperly designed structure. Failure to design for all probable loading conditions can result in an unsafe structure. Incomplete material investigations or testing and adoption of erroneous engineering properties during design can result in the use of unsuitable material in the dam or appurtenant structures. Incomplete or inadequate foundation investigations can lead to inaccurate design assumptions and improperly designed foundation treatment. Failure to recognize potential problems during the design stage can also lead to serious accidents or failure.

Defective or inferior materials used in the construction of a concrete dam can cause deterioration and possible failure of the structure. Poorly bonded cement, weak aggregates or mineral-laden water can produce low-strength concrete. Aggregates contaminated by soil, salts, mica or organic material may also produce substandard concrete. Karun Dam (Iran), Maneri Dam (India) and Kinzua Dam (USA) are some concrete dams to be worst hit due to design/construction deficiencies.

Underestimation of design floods is another factor for distress of a large number of structures. In Periyar Dam (India), three new additional vents of size 12.19m × 4.88m in continuation pf and in line with the existing regulator were provided. In Dantiwada Dam. (India), additional spillway consisting of a 308.45m-long masonry spillway with retaining walls and abutments on either side and approach channel on upstream and tail channel on the downstream side, were provided.

Erosive Forces

Erosion of concrete in hydraulic structures may occur as a result of abrasive action, cavitation or impact of impingement. In abrasion erosion, the relative motion of solid particles is nearly parallel to the concrete surface. Cavitation erosion signifies wear of concrete surface in the region of collapsing vapour bubbles in the moving water, resulting in high impact repeated collapse of cavities that cause pitting of the surface. In impact or impingement erosion, the relative motion of solid particles is nearly normal to the concrete surface. Abrasion erosion is readily recognized by the smooth, worn-appearing concrete surface, which is distinguished from the small holes and pits formed by cavitation erosion.

Erosion by abrasion results from abrasive action of water-borne silt, sand, gravel, rocks, boulders and other debris passing and/or being circulated over a concrete surface during operation of a hydraulic structure. Espinosa Irrigation Diversion Dam (USA), Kinzua Dam (USA), Nolin Lake Dam (USA), Pamona Dam (USA), Providence-Millville Diversion Structure (USA), Red Rock Dam (USA), Kentucky Dam (USA), Tarbela Dam (Pakistan), Kamafuli Dam (Bangladesh), Ilha Solteira Dam (Brazil), Marimbondo Dam (Brazil), Bhakra Dam (India), Maneri Dam (India), and Ichari Dam (India) are some dams that have experienced such erosion.

Dams that have suffered cavitation erosion are Dworshak Dam (USA), Glen Canyon Dam (USA), Lower Monumental Dam (USA), Lucky Peak Dam (USA), Yellowtail Afterbay Dam (USA), Yellowtail Dam (USA), Keenleyside Dam (USA), Hoover Dam (USA), Karon Dam (Iran), Guri Dam (Venezuela), Jupia Dam (Brazil) and Tarbela Dam (Pakistan), among others.

Large-sized boulders (size up to $0.31 m^3$) carried by Himalayan rivers, especially during floods, have been known to cause extensive damage to the concrete surface due to impact. Typical examples are low dams in the Himalayas such as Ichari Dam (India) and Maneri Dam (India).

Alkali-Aggregate Reaction

Severe distress to the body of the dam may result in the form of cracks/ bulging, etc., consequent to a chemical reaction—the Alkali-Aggregate Reaction. The most common deleterious reaction—Alkali Silica Reaction (ASR) is a reaction between dissolved alkalis (derived from cement) and active silica (in aggregates)—the other being Alkali-Carbonate Reaction, between some reactive carbonate rocks (dolomitic limestone) and alkalis in the cement. The presence of strained quartz showing undulose extinction in some rocks such as granites, charnockites and quartzites is also capable of bringing about deleterious ASR. Distress in concrete due to ASR can be reliably diagnosed on the basis of site inspection, sampling of concrete and laboratory tests and examination such as petrographic examination, scanning electron microscopy, infrared spectroscopy and concrete core tests.

Several concrete dams in the world have suffered alkali-aggregate reaction. A few examples are Churchill Dam (South Africa), Poortjieskloof Dam (South Africa), Salas Dam (Spain), Parker Dam (USA), Friant Dam (USA), Matilija Dam (USA), Stewart Mountain Dam (USA), Martin Dam (USA), Moxoto Dam (Brazil), Tarbela Dam (Pakistan), Chambon Dam (France), Shellmouth Dam (Canada), Hirakud Dam (India) and Rihand Dam (India).

Typically, alkali-aggregate reaction is evidenced by the upstream movement of an arch crown, by spalling of the concrete at extremities and by characteristic pattern cracking and crazing of the dam faces. This deleterious reaction sometimes causes the disbanding of blocks at lift surfaces. Loss of strength by disbonding, and the accompanying increase in hydrostatic pressure along the lift surfaces will reduce resistance to sliding and overturning. Alkali aggregate reaction can cause expansion of a concrete dam with consequent cracking and deterioration, and possible binding of gates, valves and metal work. Once the alkali-aggregate reactivity has developed in a relatively thin concrete dam, it cannot be stopped practically by any means. Where deterioration has progressed to a dangerously advanced stage, the effective remedies are to remove and replace the defective concrete or to build a new dam to entirely replace the old one.

Nature's Fury—Earthquakes/High Floods/Volcanic Activity

Natural disasters such as earthquakes or high floods may result in distressing the concrete dams facing the forces generated by these calamities. The response of concrete dams may by significantly influenced by variations in the intensity and characteristics of the ground motion over the width and height of the canyon; by the earthquake-induced motion of the impounded water; by the deformability of the foundation rock; and by the interactions of the motions of the water, foundation rock and the dam itself. During intense earthquake motions, vertical construction joints may slip or open; the concrete may crack; and the stored water may locally separate from the upstream face of the dam, resulting in cavitation. The filling of a large reservoir behind a high dam may actuate an earthquake. Various factors can be contributory to such a movement, including the superimposed water weight, reduction of frictional resistance in the underling rock due to pore pressures, and decline in rock strength caused by chemical alteration. It has been generally observed that large dams with a height of 100m or more show some amount of induced seismicity. A notable example in this regard is the Koyna Dam (India) (a 103m-high rubble-concrete gravity structure), where on 11 December 1967, an earthquake of Richter Magnitude 6.5 occurred, with its epicentre in the vicinity of the dam. The dam was overstressed by the earthquake motion and was damaged to an alarming degree.

Heavy floods initiate erosion action on concrete surfaces, by and large similar to erosive forces, but on a larger magnitude. Bhakra Dam (India), Ichari Dam (India) and Maneri Dam (India) are some instances where the flood fury has played a predominant role in distressing the structures.

Volcanic hazards exist in close proximity to many dams in various parts of the world. On account of volcanic explosions, large lava or mudflows can enter a reservoir, displacing the waters over a dam. Fortunately, volcanic eruptions are normally preceded by a period of associated earthquake activity and a reservoir can be lowered to increase capacity for mudflows and lava. Adequate instrumentation and seismic monitoring are essential to detect if a volcanic eruption is imminent.

Exposure of Dam Concrete to Aggressive Waters

The extent of chemical attack on dam concrete from the environment can vary considerably. In some cases, the prolonged exposure of a concrete face to any aggressive agent may simply result in unsightly surface roughness which mayor may not require remedial treatment. In the most severe cases, the main body of the dam concrete may be attacked by leaching agents, leading to the formation of seepage paths, unacceptable drainage flows and uplift pressures which give cause for structural concern. There have been cases where concrete dams affected in this way have had to be propped by earth embankments constructed immediately downstream or, in some extreme cases, abandoned completely.

Types of attack on dam concrete due to aggressive waters are indicated below:

(i) Soft water
(ii) Sulphates
(iii) Sulphides
(iv) Chlorides
(v) Acids
(vi) Plant growth
(vii) Combined attack

Soft Water—Leaching

Pure water from mountaineous regions with low dissolved salts content, having natural pH value of about 7, is aggressive to concrete and will tend to dissolve the calcium hydroxide in the set cement. This phenomenon is called leaching of lime. The hydrated calcium silicates and calcium aluminates decompose to provide further calcium hydroxide. Eventually, all the hardened cement can be decomposed in this way, leaving only a residue of aggregate and hydrates of silica, iron oxide and alumna. When the water can pass through the concrete, typically at lift or construction joints, the water evaporates, leaving an unsightly deposit of calcium carbonate on the face of the concrete, often called efflorescence.

Effects of leaching are as follows:

- Increase in seepage/leakage;
- Increase in drainage flows;
- Increase in uplift; and
- Loss of strength.

More than 20% of lime in the cement content of the concrete (on weight basis) being leached out signifies a crisis situation, necessitating urgent remedial measures. Leaching leads to choking of porous drains in concrete dams. Talkalale Dam (India), Hidkal (India), Nagarjunasagar Dam (India) and Sree Ramasagar Dam (India) are examples of some dams in India affected by leaching. Soft water attack has been reported in Arno Dam (Italy), Asswan Dam (Egypt), Avon Dam (Australia), Caspe Dam (Spain), Delta Dam (USA), Bangala Dam (Zimbabwe) and Agger Dam (Germany) apart from several others in different countries.

Sulphates

Deterioration of cement concrete by sulphate attack results by the following three alternative processes:

(i) Conversion of calcium hydroxide in the set cement to hydrated calcium sulphate and the crystallization of this compound, with consequent expansion and disruption.

(ii) Conversion of hydrated calcium aluminates and ferrites to hydrated calcium sulpho-aluminates and sulpho-ferrites with consequent expansion and disruption.

(iii) Decomposition of hydrated calcium silicates with resulting serious loss of strength.

The three principal sulphates liable to be encountered naturally are:

(i) Calcium sulphate (Gypsum or selenite)

(ii) Sodium sulphate (Glauber's salt)
(iii) Magnesium sulphate (Epsom salt)

Arranged in descending order of aggressiveness, sulphates which attack cement concrete are those of ammonium, magnesium, sodium and calcium. The most aggressive, ammonium sulphate, is not normally found in natural water. The second most aggressive, magnesium sulphate, is found primarily in seawater.

No concrete dam in the country has suffered sulphate attack of the types discussed above. Drum Afterbay dam (USA), and Toggia dam (Italy) are two typical examples of concrete dams subjected to sulphate attack.

Sulphides

Sulphides are potentially dangerous where they are able to oxidize to sulphates and free sulphuric acid.

Chlorides

Chloride attack is unusual in dam concrete. However, chloride ion concentrations greater than 0.15% by mass of cement can lead to corrosion of reinforcement in reinforced cement concrete. As per IS: 456–2000, the total amount of chloride content (as CI) in the concrete at the time of placing should not be more than 0.6Kg/m^3 of concrete (for reinforced concrete or plain concrete containing embedded metal).

Acids

Liquids containing free acids have a dissolving effect on cement paste and on aggregates containing carbonate. Various types of acids and their sources are as follows:

Tille of Acid	Sou]rce
Three mineral acids (H_2SO_4, HCl & HNO_3)	Industrial wastewater.
Carbonic acid	Rainwater dissolving CO_2 during its passage through the air.
Sulphuric acid	(i) Oxidation of sulphides such as pyrites, marcasites, pyrrhotites, chalcopyrites, etc. (ii) H_2S in degraded inundated vegetable matter.
Acid rain	Sulphur dioxide in combustion gases.
Organic acids	Decay of vegetable matter, peat, etc.

Pandoh Dam (H:P.) has suffered from sulphate attack due to the presence of pyrrhotite and chalcopyrite in the foundation rocks. Mount

Morris (USA) is a typical example of a concrete gravity dam subjected to sulphuric acid (from H_2S + bacteria) attack.

Plant Growth

Concrete surfaces can be severely affected by the growth of moss, lichens and algae. Moss roots eat into the concrete surface beneath, leaving what may originally be a hard surface, open to further attack from the environment. These unsightly growths should be removed from concrete surface during routine inspection and maintenance operations.

Combined Attack

Field conditions may combine two or more of the forms of attack described in the preceding paragraphs and, in such circumstances, the effect is generally more severe than for anyone of the forms alone.

Seepage/leakage

The main source of seepage within a concrete dam is through contraction joints or along un-bonded construction joints or lift lines. Cracks in the mass concrete are also a potential source of seepage in the structure. Uncontrolled seepage through the abutment of foundation of a concrete dam can form "pipes" or voids, causing the bridging of sections of the abutment and resulting in concentration of the stresses in the concrete.

Many gravity dams constructed in the nineteenth century were of stone masonry in lime mortar, which is susceptible to deterioration and loss of strength over long periods of exposure to seeping water. Once the bond is broken, water pressure in the joints may actuate a sliding or overturning failure.

Prominent examples of rubble concrete/concrete dams adversely affected by seepage/leakage include Koyna dam (India), Hirakud dam (India) Nagarjunasagar dam (India) and Periyar dam (India). Examples of masonry dams include Barna dam (India), Talkalale dam (India), Donkarayi dam, Hemavathy dam (India), Bhadra dam (India), Bargi dam (India), Kadana dam (India), Parbati dam (India) and Tigra dam (India).

Rebar Corrosion

The corrosion of steel in concrete has been found to be a major factor affecting the durability of reinforced concrete structures in a project. The corrosion process is largely a result of individual or combined action of factors such as carbonation, chloride contamination, influence of cracks and inadequate cover to reinforcement. Though appropriate investigations, design and execution measures such as absence of harmful constituents, adequate cover thickness, good quality concrete, impermeability, etc., it is possible to protect the reinforced cement concrete structures even under aggressive environment. The corrosion, once triggered, requires appropriated remedial measures for rehabilitation of the distressed structures.

Freezing and Thawing

Concrete mixes for massive structures usually contain an air-entering agent. This appreciably improves the durability of the concrete and increases resistance to freezing and thawing. However, such distress can still occur where air entrainment is insufficient or when the coarse aggregate is itself vulnerable to freeze-thaw action. Closely spaced parallel cracks at edges of concrete blocks may be symptomatic of freeze-thaw expansion. Entrance of water into the cracks and subsequent freezing are likely to further the deterioration. Freeze-thaw effects are most likely to be found in parapets, cantilever beams, slabs and walls of appurtenant structures.

Ageing of Concrete

Ageing of concrete dams can be attributed to both physical and chemical factors. The former relate to changes in forces acting on the structures, including those caused by temperature variations, and deterioration due to alternate wetting and drying. The latter are associated with infiltration, into the dam of either pure (soft) water or aggressive waters, alkali-aggregate reaction, etc.

Waterstop Failures

Joints are a necessary part of any large construction where dimensional changes within the structure are likely to occur. This is particularly true of concrete dams where, as a result of the nature and behaviour of mass concrete, they may suffer cracking due to a greater or lesser degree unless joints are provided to control the various problems. The various types of joints are horizontal lift joint, construction joint, contraction joint, expansion joint, isolation joint, control joint and cold joint. To avoid or limit the passage of water along the joints, it is necessary to provide a seal across the joint in order to prevent excessive flow of water along the seam. Such seals must be strong enough to withstand rough treatment during construction and the head of water, flexible enough to accommodate relative movement between the monoliths and durable enough to be effective during the lifetime of the structure, which in concrete dams, can be more than 100 years. The various materials which may be used for joint sealing include polyvinyl chloride (PVC), natural or synthetic rubber, copper, stainless steel and bituminous compounds. Experience shows that the sheet type copper waterstop is very vulnerable to failure. Effective repairs to monolith joint waterstops can be achieved by various methods listed below in concrete dams:

- Surface plates backed with a deformable pad.
- Sawed joint filled with an elastic sealant.
- Injecting chemical grout through drilled holes that intercept the joint.
- Drilling a hole along the joint and filling it with elastic chemical grout.
- The particular method to be used depends on a number of factors including the degree of joint openness and movement, height of the joint, environment, physical abuse, economics, available construction time and access to the upstream joint face.

EMBANKMENT DAMS

- The reasons for distress in an embankment dam may be one or more of the following and the distress may manifest itself in the form of cracks, settlement, displacement, sinkholes, wet spots, boils, surface erosion, animal burrows, vegetation, depressions, slides, deterioration of slope protection, bulging, openings on the embankment crest or slope, etc.
- Foundation failure caused due to erosion, solution, settlement, etc., resulting in settlement of embankment followed by cracks.
- Foundations with low shear strength or with seams of weak material.
- Seepage through dam or its foundation.
- Inadequate investigations of site and fill materials, leading to the use of materials such as erodible soils, dispensive soils, expansive soils or placing foundations on collapsible soils, without specific tests.
- Overtopping caused due to inadequate spillway capacity or non-functioning of flood gates. Embankments may also be susceptible to erosion unless protected from wave action on the upstream side and surface run off on the downstream face.
- Internal erosion causing piping or movement of material at the dam foundation interface or along the contact of embankment with adjoining or embedded structure. The presence of gap-graded materials such as open work gravels.

- Ageing of dam with time-related changes in properties of materials.
- Liquefaction of cohesionless material in foundations.
- Adverse conditions which have been well known in respect of embankment dams and which merit attention, are listed below:
  - Poorly sealed foundation
  - Cracking in the core zone.
  - Cracking at zonal interfaces.
  - Soluble foundation rock
  - Deteriorating impervious structural membranes
  - Inadequate foundation cut offs
  - Desiccation of the clay fill
  - Steep slopes vulnerable to sliding
  - Blocky foundation rock susceptible to differential settlement
  - Ineffective contact at adjoining structures and at abutments
  - Pervious embankment strata
  - Vulnerability to "quick" conditions during an earthquake. Hydraulic fill dams are especially suspect in this respect.
  - Distortion at critical points due to poor compaction and too low moisture content at the time of placement of embankment material.
  - Sliding of reservoir slopes or the dam abutments.

The failure of a dam may be caused due to anyone or more than one cause and identification of these is the sole purpose of investigation. However, it may not be possible to identify the main factor initiating instability and prepare a sequential chart of factors leading to catastrophic consequences. Time and finances are the main hurdles but approach should be so as to reach the goal without any bias. An incomplete report is normally the end result—intentionally or otherwise—without adding to knowledge, as stated above.

OVERTOPPING OF DAMS/CONTROVERSIAL FAILURES

One of the common causes of failure of dams is attributed to 'unprecedented rains in region', which is considered as a 'summary disposal' of failure case. It is interesting to note that this one reason, when applied to all types of dams gives different meanings; the primary reason may be hydrological or adequate spillway resulting in overtopping. For instance, the overtopping of a composite dam, the Kaddam dam (1958), caused erosion of earth backing and the thin masonry section gave way, resulting in failure of the dam. The earth dam, Panchet dam (1961), developed piping conditions and in this state, overtopping resulted in failure of the dam. The overtopping caused breach of a saddle dam in the case of composite dam, Dantiwad dam (1973). Kodganar Earth dam failed in 1977, due to unprecedented rain in its catchment. The dam was overtopped and it is reported that the entire length of the dam overflowed for an hour, before four breaches occurred. The main reason was the inability to lift spillway shutters due to power failure. The diesel generators also failed, forcing the use of manual operation. About three to four hours were lost, causing overtopping and consequential damages. The body of a *concrete dam* may sustain overtopping, provided measures to save toe erosion are taken. Relatively, a *'flow-over' type of embankment dam'* is difficult to design. ICOLD (1985), Q.59, includes examples of Bearspaw dam in Canada, Henshaw dam in the USA, Loerie dam in South Africa, Stanford dam in Great Britain and numerous dams in the USSR.

Casagrande, Wilson and Schwantes (1972) Contrary to common impression that Baldwin Hills Dam failed, emphasized that the dam itself performed satisfactorily during the entire life of reservoir and did not contribute to the failure of reservoir. It is questioned that 'to what extent the subsidence, accompanied by tensile strains at the Baldwin reservoir site' were caused due to oil recovery or by tectonic movements. It remained a controversial question. Teton Dam failed in 1976, during its first filling. It had controversial technical reasoning, as included in the Case Studies. But, according to Penman (1986) Teton site was 'suitable for a reservoir but not for a dam'. Penman also felt that un-instrumented dam could not have stopped the failure but 'would have taught the engineers a lot about the disaster subsequently'.

INVESTIGATION OF FAILURES

To investigate causes of a failure, it is necessary that before getting biased, a thorough inspection of site conditions is made at the earliest opportunity after the failure and as many times as necessary, so as to visualize the incident from the post-failure conditions. It may be useful that a group, with engineers of varied experiences and expertise in dam design, construction and surveillance/failure, is formed and works independently till a periodic meeting for group discussion is necessary to take a comprehensive view of the failure and finalize the report.

The following steps are suggested without any significance to the order in which these are mentioned:

1. Collect information about sudden changes in environment if any, such as occurrence of unprecedented rain, flash floods, earthquake shocks, etc., prior to failure.
2. To note changes, if any, in operation of reservoir, either due to changes in demand of water or due to obvious changes in performance of dam before the failure.

3. Directly contact the victims or sufferers in the neighborhood, who have witnessed the failure and will give sequence of events, prior to failure.
4. Collect site investigation reports, design and construction reports, drawings, etc., including records of deviations from the designed cross section and foundations during construction with reasons.
5. Keep track of quality control and quality assurance records of different reaches of dam. Back analysis, if any, may be carried out with quality control records, utilizing actual properties of materials used in the structure and foundation.
6. Performance records, surveillance reports supported by instrumentation data, if available and the unexpected or different behaviour of dam than that predicted.
7. Risk Analysis if carried out. How the actual failure tallies with the predicted potential mode of failure and sequences of events. Johnson and Illes (1976) give the probable shapes of breaches after dam failure. The intact portions of dam on the sides give the extent of the damaged length of the dam, its shape, crunching or cutting through foundations.
8. Investigate the properties of materials in the intact portion of the dam and its foundation.
9. Analyse aerial photographs/satellite imagery before, during and after the failure of a dam.
10. There are a number of techniques for in situ evaluation of overall quality of concrete in concrete dams. The most commonly used techniques are as follows:
    - (i) Surface harness method—Schmidt Rebound Test Hammer.
    - (ii) Ultrasonic pulse velocity technique—Portable Ultrasonic Non-destructive Digital Indicating Tester (PUNDIT).
    - (iii) Radar survey—Ground penetrating radar.
    - (iv) Sonic methods (sonic log and sonic crosshole).
    - (v) Sonic Tomography.
    - (vi) Detection of reinforcement in concrete Digicover or Pachometer (Covermeter).
    - (vii) Detection/Monitoring of Corrosion of Reinforcement:
        - (a) Corrosion Potential Measurements:
            - Potential Wheel and Data Bucket System.
            - Colebrand Pathfinder.
        - (b) Resistivity of concrete.
        - (c) Electrical resistance measurements.
        - (d) Infrared thermography

(viii) Underwater surface scanning using Remote Controlled Underwater Vehicle (ROV).
The water scanning system called 'Remote Operated Underwater Vehicle (ROV) comprises a heavy submersible vehicle, control cosole, cameras, a VCR and other accessories. The vehicle, along with the camera, can be lowered into the reservoir water upto a depth of 300 metres and a video film can be made showing the dam surface and other features. The video film can be studied subsequently to assess the extent of damage so that possible remedial measures can be taken up.

11. A number of techniques for in situ evaluation of the overall quality of embankment material and foundations are available, which can be adopted as per requirement. These are briefly mentioned below:
    (i) Collection of undisturbed and disturbed soil samples for characterisation of the material in the embankment or/and foundation. The laboratory tests should be tailored to confirm or rule out the possible suspected problem.
    (ii) Field tests for characterisation of permeability, embankment material and foundation strata for in situ shear strength.
    (iii) Tracer techniques (Dye or nuclear) for seepage detection/ mapping.
    (iv) Sonic tomography.
    (v) Radar survey—Ground penetrating radar.
    (vi) Seismic refraction and reflection.
    (vii) Seismic shear wave velocity investigations—Downhole, Crosshole and Uphole.
    (viii) Resistivity surveys.
    (ix) Nuclear density and moisture gauge.

The above data, as collected and analysed, may lead to a number of lessons and the reviewer—even forming a group—may agree partly or differ in conclusions. This difference may be due to difference in experiences or bias to the subject of specialization, as presented in the case histories with controversial reasons of failures. However, it may also be kept in view that it is not necessary that all the failures case studies need fitting into well-known principles of behaviour of materials and structures, under a particular environment. An engineer's or scientist's knowledge may be limited but not the nature, which also extends beyond the realm of human understanding, the *knowledge* yet to be explored or *experience* yet to be acquired. Open-minded political reasons, not free from arbitrary aspects, also need consideration: *'What is, for mankind, the price of a human life?'*

APPENDIX

STRUCTURES, FEATURES, EVENTS AND EVIDENCE TO BE EXAMINED DURING INSPECTION

1. GENERAL

The performance of dams and appurtenant structures is controlled by:

(1) their designs,
(2) the characteristics of their constituent materials,
(3) the nature of their foundations, and
(4) their regional settings.

The fundamental objective of dam safety inspections is the detection of any existing or developing structure of hydraulic weakness. In searching for weaknesses, dam inspectors must recognize and understand the interrelationship of the demands which control the performance.

2. CHANGES IN THE CHARACTERISTICS OF MATERIALS

2.1 General

Observe for defective, inferior, unsuited, or deteriorated materials. A variety of different materials make up the different types of dams and appurtenances. The quality and durability of these materials must always be determined in every instance and the need for such critical examination and what to look for as listed in this section are not normally repeated in the sections on specific structures.

2.2 Concrete

(1) alkali-aggregate reaction, pattern of crushing and cracking
(2) leaching
(3) abrasion
(4) spalling
(5) general deterioration
(6) strength loss

2.3 Rock

(1) degradation
(2) softening
(3) dissolution

2.4 Soils

(1) degradation
(2) dissolution
(3) loss of plasticity
(4) strength loss
(5) mineralogical change

2.5 Soil-cement

(1) loss of cementation
(2) crumbling

2.6 Metals

(1) loss of cementation
(2) corrosion
(3) stress-corrosion
(4) fatigue
(5) tearing and rupture
(6) galling

2.7 Timber

(1) rotting
(2) shrinkage
(3) combustion
(4) attack by organisms

2.8 Rubber

(1) hardening
(2) loss of elasticity
(3) heat deterioration
(4) chemical degradation

2.9 Joint Sealers

(1) loss of plasticity
(2) shrinkage
(3) melting

3. GENERIC OCCURRENCE

3.1 General

Expects should study and observe occurrences for their characteristics, locations and recency. These occurrences are of a universal nature regardless of the structure type of foundation class. The details of what to look for in observing these occurrences, actual of evidential, are not repeated in the sections on specific structures.

3.2 Seepage and Leakage

(1) discharge-tag relationship
(2) increasing or decreasing
(3) turbidity and piping
(4) colour
(5) dissolved solids
(6) location and pattern
(7) temperature
(8) taste
(9) evidence of pressure
(10) boils
(11) recency and duration

3.3 Drainage

(1) obstructions
(2) chemical precipitates and deposits
(3) unimpeded cutfall
(4) sump pump facilities
(5) bacterial growth

3.4 Cavitation

(1) surface pitting
(2) sonic evidence
(3) vapour pockets

3.5 Stress and Strain: Evidence and Clues

(1) in concrete—cracks, crushing, displacements, offsets, shears, creep,
(2) in steel—cracks, extensions, contractions, bending, buckling.
(3) in timber crushing, blocking, bending, shears, extensions, compressions
(4) in rock and soils—cracks, displacements, settlement, consolidation, subsidence, compression, zones of extension and compression.

3.6 Stability, Evidence and Clues

(1) in concrete and steel structures—tilting, tipping, sliding, overturning
(2) in embankment structures, cut slopes natural slopes—bulging, sloughing, slumping, sliding, cracks, escarpments
(3) in rock cutslopes, foundation and unlined tunnels—slumps, slides, rockfalls, bulges, cracks.

4. OPERATION AND MAINTENANCE

4.1 Service Reliability of Outlet, Soilway Sump Pump Mechanical-electrical Features:

(1) broken or disconnected lift chains and cables
(2) test operation including auxiliary power sources
(3) reliability and service connections of primary sources
(4) verification of operators understanding and ability to operate
(5) case and assurance of access to control stations
(6) functioning of lubrication systems

4.2 Gate Chamber, Galleries, Tunnels and Conduit

(1) ventilation and heat control of damp, corrosive environment of mechanical-electrical equipment.

4.3 Accessibility and Visibility

(1) obscuring, vegetal overgrowth
(2) galleries—access ladders, lighting
(3) access roads and bridges
(4) communication and remote control lines, cables and telemetering systems

4.4 Control of Vegetation and Burrowing Animals

(1) harmful vegetation in embankments—oversize, dead root channels
(2) harmful vegetation in structural concrete joints
(3) obstructing vegetal growth in hydraulic flow channels
(4) ground squirrels, rats and beavers

5. BEHAVIOUR

5.1 Every attempt should be made to anticipate and have engineer-observers present on site at items of large spillway and outlet discharge.

Resident operational personnel can often supply valuable information and may be the only available observers (during earthquakes) for example.

5.2 Warning, Safety and Performance Instrumentation

(1) piezometers, flow recorders, accelerometers, seismoscopes, joint meters, and gauge points, strain meters, stress meters, inclinometers, direct and inverted plumblines, surface reference monuments, stage recorders, extensometres,
(2) serviceability
(3) access to readout stations
(4) type and location suitable for condition being observed
(5) need for recalibration
(6) faulty readings, sources and reasons
(7) alarm systems operable and at appropriate set points
(8) random check readings during inspections
(9) quiz operators to determine their understanding of purpose and functioning of instruments

5.3 During and After Large Floods

(1) drift-marked high water-lines
(2) evidence of taxed spillway capacity
(3) undesirable or dangerous spillway flow patterns directly observed or deduced from flow strains, erosion trails, swept vegetation, deposition of solids

5.4 During and After Large Outlet Releases

(1) undesirable or dangerous spillway flow patterns, dynamic pressures, vibrations cavitation sonics.

5.5 After Earthquakes

(1) cracks, displacements offsets in structural features
(2) cracks, slumps, slides, displacements, settlements in embankments, cut-slopes, and fill slopes
(3) broken stalactites in galleries, tunnels, chambers
(4) toppled mechanical equipment
(5) sand boils

6. CONCRETE DAMS

Any of these observations are also applicable to reservoir impounding power intake structures, spillway control structures, lock walls.

6.1 Stress and Strain

(1) cracks, crushing, displacements, offsets in concrete monoliths, buttresses, face slabs, arch barrels visible on exterior surfaces and in galleries, value and operating chambers, and conduit interior surfaces,
(2) typical stress and temperature crack patterns in buttresses, pilasters, diaphragms and arch barrels
(3) retention of design forces in post-tensioned anchorages and tendons.

6.2 Stability

(1) excessive or maldistributed uplift pressures revealed by piezometres, pressure spurts from foundation drain holes, construction joints, and cracks
(2) differential displacements of adjacent monoliths, buttresses and supported arch barrels or face slabs
(3) disparities in region near the interface between arches and thrust blocks
(4) movement along construction joints
(5) uplift on horizontal surfaces revealed by seepage on down stream face and in galleries at construction lift elevations.

6.3 Hillsides and river channels adjacent to the abutments and river section foundation along the downstream toe of the dam

(1) leakage
(2) seepage
(3) stability
(4) boils

6.4 Special attention to stability and seepage control at discontinuities and juncture

(1) embankment wrap around sections
(2) waterstops in monoliths and face slabs
(3) reservoir impounding backfill at spillway control sections and retaining walls

6.5 Foundation

(1) piping of weathering products from old solution channels and rock joint structure
(2) efficiency of foundation seepage control systems—drains, drainage holes, grout curtains, cutoffs, drainage tunnels

(3) history of shear zones, faults, cavernous openings
(4) zones of varying permeability
(5) orientation of stratification and bedding planes—effect on permeability, uplift, foundation stability
(6) subsurface erosion and piping
(7) thin weaker interbeds—effect on stability

7. EMBANKMENT-TYPE DAMS

7.1 Evidence and Clues: Stress and Strain

(1) settlement
(2) consolidation
(3) subsidence
(4) compressibility
(5) cracks, displacements, offsets, joint opening changes in concrete facing on rockfills
(6) loss of freeboard from settlement
(7) zones of extension and compression visible along dam crest or elsewhere
(8) crushing of rock points of contract
(9) differential settlement of embankment cross sectional zones visible along the dam crest, indicating stress transfer along the region of zone interface (increases possibility of hydraulic fracturing)

7.2 Stability: Evidence and Clues

(1) cracks, displacements openings, offsets, sloughs, slides, bulges, escarpments on embankment crest and slopes and on hillsides adjacent to abutments,
(2) sags and misalignments in parapet walls, quardrails, longitudinal conduits or other linaments parallel to embankment axis.
(3) irregularities in alignment and variances from smooth, uniform face planes
(4) bulges in ground surfaces beyond toes of slopes

7.3 Evidence and Clues

(1) wet spots
(2) new vegetal growth
(3) seepage and leakage
(4) boils
(5) saturation patterns on slopes, hillsides and in the streambed

(6) depressions and sinkholes
(7) evidence of high escape gradients

7.4 Erosion Control

(1) loss, displacement, and deterioration of upstream face riprap, underlayment and downstream face slope protection
(2) leaching

7.5 Foundation

(1) see 6.5 –also
(2) consolidation
(3) liquefaction

7.6 Other endangerments

(1) utility pressure conduct on, over, or through embankments
(2) diversion ditches along abutment hillsides

8. SPILLWAYS

8.1 Approach Channel

(1) obstructions
(2) slides slumps and cracks and cutslopes

8.2 Log Booms

(1) submergence
(2) uncleared accumulated drift
(3) parting
(4) loss of anchorage
(5) inadequate sleek for low reservoir stages

8.3 Hydraulic Control Structure

(1) stability
(2) retention of capacity rating
(3) erosion at toe
(4) unauthorized installations on crest, raising storage level and decreasing spilling capacity
(5) gate piers
(6) trash control systems

(7) nappe and crotch aeration
(8) siphon prime settings

8.4 Headwater Control (Gates, Flashboards, Fuse Plugs)

(1) unauthorised position
(2) wedging
(3) gate trunnion displacements
(4) loss of gate anchorage post-tensioning
(5) undesirable eccentric loads from variable positions of adjacent gates
(6) gate-seal
(7) erosive seal leakage
(8) failure of system
(9) availability of bulkhead facilities for unwatering and of cranes and lefting beams

8.5 Operating Deck and Hoists

(1) broken or disconnected lift chains and cable
(2) unprotected exposure of electrical-mechanical equipment to weather, sabotage, vandalism
(3) structural members and connections

8.6 Shafts, Conduits, and Tunnels

(1) vulnerability to obstruction
(2) evidence of excessive external overloading pressure jets contorted cross-sections, cracks, displacements, circumferential joints
(3) serviceability of linings (concrete and steel), materials deterioration, cavitation, erosions
(4) rockfalls
(5) severe leakage about tunnel plugs
(6) support system for pressure conduits in walk in tunnels

8.7 Bridges

(1) possibility of collapse with consequent flow obstruction
(2) serviceability of operational and emergency equipment transport

8.8 Discharge Carrier (Open Channel or Conduit)

(1) vulnerability to obstruction
(2) evidence of excessive external sidewall loading—large wall deflections, cracks, differential deflections at vertical joints

(3) invert anchorage and foundation support—dummy sounding, buckled lining, excessive uplift
(4) observation of evidence of dangerous hydraulic flow patterns cross water, inadequate free board, wall climb, unwetted surfaces, uneven distribution, ride up on horizontal curves, negative pressures at vertical curves, pressure flow deposition,
(5) drain system serviceable
(6) air injection and expulsion
(7) tendency for jump formation in conduits
(8) buckling, slipping of slope lining
(9) erosion of unlined channels

8.9 Terminal Structures

(1) inadequate dissipation of energy
(2) jump sweepout
(3) undercutting
(4) retrogressive erosion
(5) loss of foundation support for flip bucket substructure
(6) unsafe jet trajectory and impingement
(7) erosive endangerment of adjacent dam or other critical structures

8.10 Return Channels

(1) impaired outfall
(2) obstruction
(3) slides, slumps, cracks in cutslopes
(4) erosion of deposition creating dangerous tailwater elevations or velocities,
(5) evidence of destructive eddy currents

9. OUTLETS

9.1 General

Many of the observations made of outlet components are similar in nature and purpose to these made for spillway components, stilling basins, for example.

9.2 Approach channels (may seldom be directly visible and may require underwater inspection)

(1) siltation
(2) underwater slides and slumps

9.3 Intake structure (including appended inclined and free-standing tower with wet and dry)

(1) lack of dead storage
(2) siltation
(3) potential for burial by slides and slumps
(4) damage or destruction of emergency and service bulkhead installation facilities
(5) availability of bulkhead, cranes, lifting beams
(6) serviceability of access bridges

9.4 Trashracks and Raking Equipment

(1) clogging of bar spacing
(2) lodged debris on horizontal surfaces
(3) collapse

9.5 Gate Chambers, Gates, Valves, Hoists, Controls, Electrical Equipments, Air Demand Ducts

(1) Accessibility to control station under all conditions
(2) Ventilation
(3) Unauthorised gate or valve positions
(4) Binding of gate seals
(5) Seizing
(6) Erosive seal leakage
(7) Failure of lubrication system
(8) Drainage are sump pump serviceability
(9) Vulnerability to flooding under reservoir pressure through conduits, by-passes and gate bonnets surfacing in the chamber

9.6 Conducts and Tunnels

(1) See 8.6 also
(2) Seepage or leakage along external periphery of conduit
(3) Extension strains in conduits extending through embankments
(4) Capacity and serviceability of air relief and vacuum valves on conduits

9.7 Terminal Structures

See 8.9

9.8 Return Channels

See 8.10

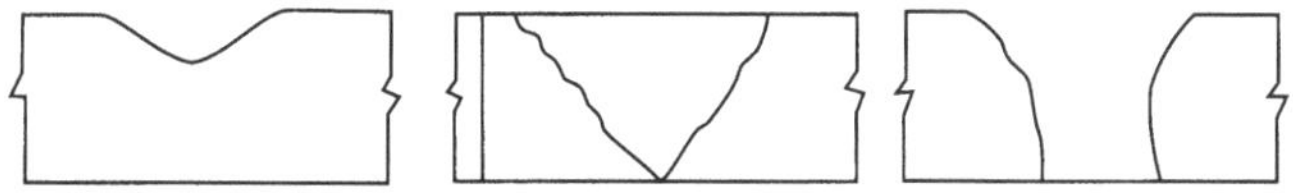

Fig. 5.1 Breach shapes resulting from over-topping, showing: (a) Initial notch shape: (b) "V" shape formed when softer dam material is eroded away: and (c) resulting lateral erosion

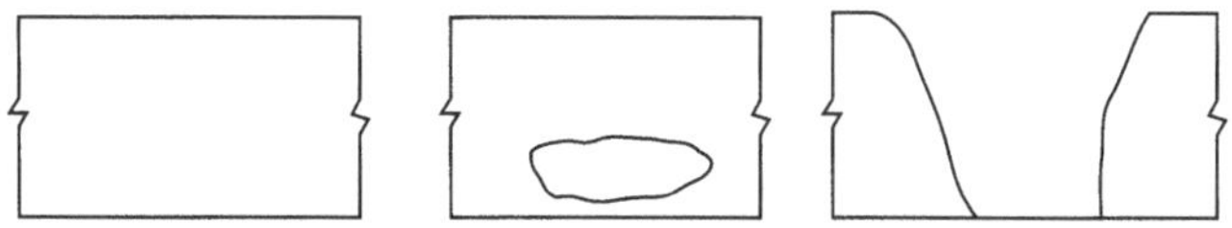

Fig. 5.2 (a, b, and c) Sequence of breach shapes resulting from a piping failure. It should be noted that the final shape is similar to the corresponding stage for an overtopping failure, as shown in 1(c)

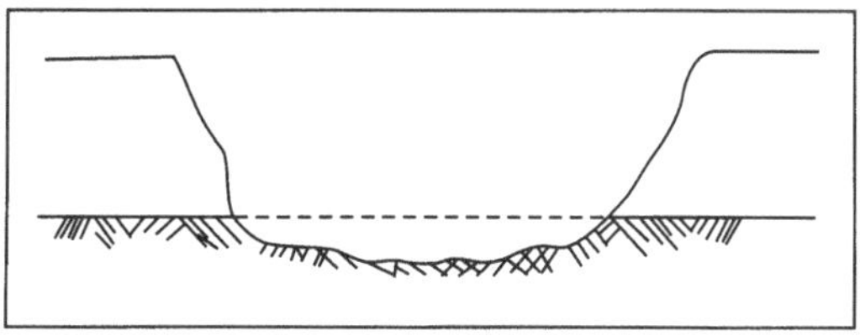

Fig. 5.3 Shape resulting from an earth dam foundation failure

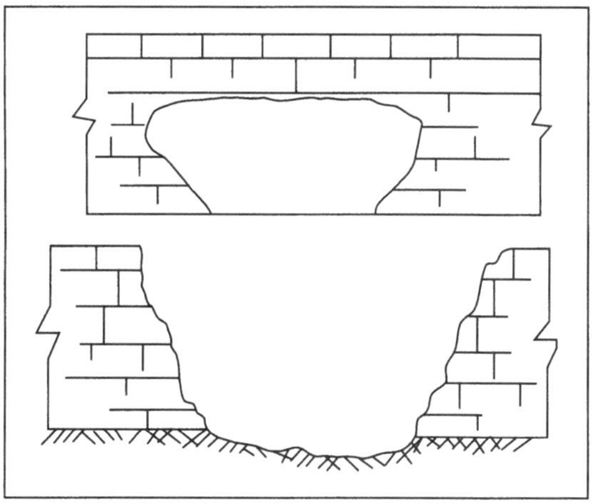

Fig. 5.4 Breach formations resulting from a gravity dam foundation failure

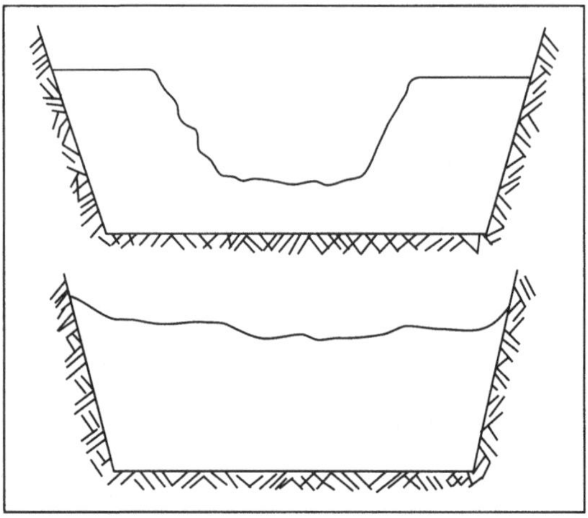

Fig. 5.5 Typical shapes resulting from gravity dam overtopping

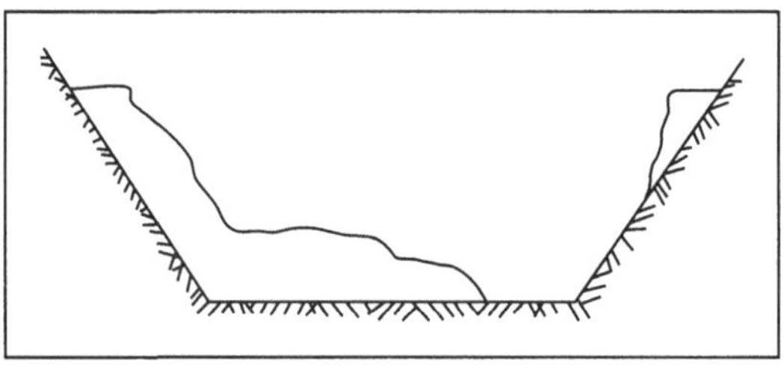

Fig. 5.6 The Malpasset failure

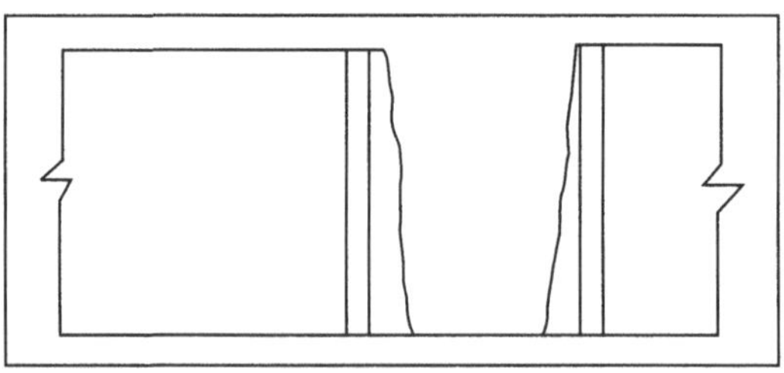

Fig. 5.7 Breach shape resulting from a buttressed arch dam failure

REFERENCES

ASCE (1988): 'Evaluation Procedures for Hydrologic Safety of Dams', *A Report by the The Task Committee on Spillway Design Flood Selection.*

Bhasin, P. C. (1982): 'Special Lecture', *Seminar on Lessons from Failure of Structures* Organized by Maharashtra India Chapter of ACI, Bombay.

Carlyle, W. J. (1984): 'Legislation on Safety of Dams', *Report to Committee on Dam Safety,* ICOLD.
Casagrande, A. (1964): 'Role of the Calculated Risk in Earthwork and Foundation Engineering', *Jr. of the Soil Mechanics and Foundation Division,* ASCE, Vol. 91, SM4, July, pp. 1-40.
Casagrande, A., Stanley D. Wilson and E. Douglas Schwantes Jr (1972): 'The Baldwin Hills Reservoir Failure in Retrospect', *ASCE Specialty Conference,* Performance of Earth-Supported Structures, Purdue University, Lafayette, Indiana, pp. 586.
ICOLD (1987): Dam Safety Guidelines, Bulletin 59.
ICOLD (1994) *Understanding the Need for Dams and their Safety.* Prepared by the Committee on Dam Safety for the International Commission on Lange Dams.
ICOLD (1994) *Understanding the Need for Dams And Their Safety,* Prepared by Committee on Dam Safety.
ICOLD (1994): '*Ageing of Dams And Appurtenant Works,* Bulletin 93, pp. 17.
Jansen, Robert B. (1980): 'Dam Safety Engineering', Section 7 of *Report of Development of Dam Engineering in the United States,* pp. 1032-1049.
Jansen, Robert B., Gordon W. Dukleth, Bernard B. Gordon, Laurence B. James and Clyde E. Shields (1967): 'Earth Movement At Baldwin Hills reservoir', *Journal of the Soil Mechanics And Foundation Division,* ASCE, vol 3, No. M4, pp. 574.
Johnson F.A. and Illes P. (1976) 'A Classification of dam failures', *Water Power & Dam Construction,* December 1976 pp. 43-46.
Lafitte, Raymond (1985): 'GR. Q. 59, Rehabilitation of Dams To Ensure Safety', *Proc. ICOLD,* Congress, Lausanne, pp. 986
Leonards, Gerald A. (1982): 'Investigation of Failures', Sixteenth Terzaghi Lecture, Journal of the Geotechnical Engineering Division, GT 2, February.
Menezes, M. (1982): 'Keynote Address', *Seminar on Lessons from Failures of Structures.* Organized by Maharashtra India Chapter of ACI, Bombay.
Penmaw, ADM (1986) 26th Rankine Lecture. On the Embankment Dams. *Geotechnique* vol. 36 no. 3 pp. 301-348.
Scott, R.F. (1987): 'Failure' 27th Rankine Lecture, *Geotechnique* 37, no. 4, pp. 423-4660.
Seed, H.B. and Duncan J.M. (1981): 'The Teton Dam Failure—A Retrospective. *Review Proceedings of the* X *International Conference on Soil Mechanics and Foundation Engineering,* Stockholm, pp. 2191.
Thatte, C. D. (1991): 'Introductory Address', *Workshop on Designing of Dam for Hydrologic Safety, Post-Session Proceedings,* CBI&P, New Delhi, pp. 4.
Water Power & Dam Construction (1985): 'Four Major Dam Failures Re-examined', *International Workshop on Dam Failures,* Purdue University, Lafayette, Indiana, USA, November, pp. 33-460.

Chapter 6

Study of Dam Failure—A Philosophy

DAM FAILURE

Failure has always been considered as a stepping stone to success. In other words, success is always enjoyed in absolute term in the background of failure. But failure of one dam wipes out the successful achievements of civil engineering. One, individual killed cannot be tolerated against thousands saved and who enjoy the basic comforts of life, drinking water, food and industrial developments due to a dam holding a water reservoir. It is rightly said that one of the excitements of engineering is that it permits an engineer to use his judgement, where knowledge fails to help him and to go ahead with design and construction. In his Terzaghi lecture, Casagrande (1964), for the first time, used the term *'Calculated Risk'*. He defined two aspects of risk

1. The use of imperfect knowledge, guided by judgement and experience
2. The decision on an appropriate margin of safety or the degree of risk

Referring to Failure of Baldwin Hills Reservoir Casagrande questioned-"— were the persuasive arguments responsible for taking grave risk or did the designers underestimate the risk?"

The complexity of engineering structure always have some problems which remain unsolved and if an engineer has to wait or search for solutions, perhaps the structure will never come up. It is the calculated risk, tempered with sound judgement, that definitely proves itself to be solution of the problem, unless an unknown situation crops up, which has remained beyond the comprehension at the design stage.

One of the reasons of dam failure—very uncommon—is the lack of coordination among the investigation, design and construction engineers. More so for the reason that the agencies responsible for these activities are normally different and even if desired by one, the others remain beyond reach. A close coordination is an important guarantee against failures. But such teamwork poses a problem during investigations of dam failure, as fixing the responsibility may create embarrassing situation. Of late, team working has become difficult as specialisation is producing a different *class of engineers,* with languages of their own. Interdisciplinary approaches and teamwork is one of the solutions to avoid risk of failure, irrespective of any reasons, primarily perhaps due to one's ego.

Human error remains one of the prime reasons of dam failure and cannot be ignored in study of failures, irrespective of sophisticated analyses adopted for design or automated and computerized techniques used in instrumentation. This is much more important in developing countries as projects are mostly planned *labour oriented* in order to provide job opportunities. At the construction of a Nagarjunasagar dam, a masonry dam in India, in three shifts of a day, 30,000, laborers were employed per shift. The other factor is transfer of advanced technology to these countries without properly trained staff at the root level, who are responsible for construction. This leaves a wide gap in theory and practice at the cost of hazardous risk. Certain failures have exposed this lacuna in dam engineering. For lack of appraisal of certain critical design features, like foundation treatment of dams, these are inadvertently ignored during construction.

Similarly, in embankment dams, the lack of knowledge of field staff about the significance of water content in soil and compaction quality has been responsible for failures due to erosion, seepage or excessive settlement of dam. By proper training, these types of errors are avoidable. But the design, construction and maintenance of dams present interacting activities, and it is difficult to eliminate errors, which is rightly said, may be *gross errors, or systematic errors or random errors.* These may be a simple mistake or a mistake in assumption or failure of a probability. The study and dissemination of the source of error is a social responsibility, but how far can this obligation towards society be fulfilled is a matter to ponder.

STUDY OF FAILURES

Dam failure is a single catastrophic event and a number of such events have occurred, but normally these are underplayed, unless loss of lives is involved. However, all said and done, it is always correct to assume that irrespective of reasons, it is never an intentional failure for which a single person or a group is responsible. The engineering concept of safety of a structure, with only change in degree, has attained a different angle these days. A dam is built with a finite risk of failure, as without this approach it is not possible to construct it. For occurrence of an event, once in a thousand years, the design of dam cannot be made totally safe. Lower the risk means an uneconomical project and cost-benefit ratio has to be neglected for which the society will not agree. Hence, construction of dam, like any other structure, is taken up under socio-economic understanding, for which an engineer alone may not be responsible.

The basic philosophy of study of failure should be that revelations as an outcome need technical considerations rather than witch-hunting. The report of dam failure should be based on logic supported by authentic facts as revealed by observations, field and laboratory testing and pre-and post-failure records. It may not be possible to make the study unbiased, as logic is based on an individual's experience and subject specialization. This is likely to lead to controversial conclusions. In a bureaucratic setup, these controversies are exploited for undesirable ends. Like in other professions, engineers, agreeing to disagree among themselves should establish certain ethics to safeguard interests of engineering community. The best way is to stress on *lessons* that are learnt by the study of failures, rather than pressing for acceptance of a particular logic. Unless this approach is adopted, studies in this field will remain incomplete and to repeat Leonards (1982) the failures neither lead to more rational design methods nor in improvement of methods of investigation of failures.

REFERENCES

Casagrande, A. (1964): 'Role of the Calculated Risk in Earthwork and Foundation Engineering', *Jr. of the Soil Mechanics and Foundation Division,* ASCE, Vol. 91, SM4, July, pp. 1-40.

Leonards G.A., (1982): 'Investigation of Failures'. Sixteenth Terzaghi Lecture. *Jr. of the Geotechnical Engineering Division,* GT 2, February.

Subject Index